BEYOND BIGFOOT & NESSIE

LESSER-KNOWN MYSTERY ANIMALS FROM AROUND THE WORLD

KATE SHAW

Strange
Animals

For Michael J.

In memory of that one perfect summer when we hiked the Andes Mountains with nothing but the clothes on our backs and a llama named Bingo. I still remember those evenings around the campfire, you playing folk songs on the cello you made from bark and llama hair, Bingo and me singing along under the stars. It was all good times until Bingo double-crossed us in La Paz.

Here's to adventures large and small in the future.

CONTENTS

FOREWORD

Congratulations, reader! You're looking at an incredibly useful book for raising monster awareness. It's a strange old world, and full of all kinds of unusual creatures. Throughout history people have been dealing with monsters in every culture humans have developed. Sometimes the monsters are real animals. Sometimes they're fantastical and mythic. But it's the borderlands between these two domains that really capture the imagination because in that shadowy realm, the strange cry you hear in the forest or the blurry shape moving in the background of your vacation photos *might just be a monster*.

The amateur science field of cryptozoology is dominated by the titans of monster media: Bigfoot, Nessie and the Abominable Snowman. A few B-level monster celebrities also get a lot of coverage: Chupacabra, Mothman, Ogopogo, Champ... But what about the thousands of monsters that haven't even started a social media account? I jest, but the point is sincere. There are many more monsters out there than you've likely ever heard of before and Kate Shaw has put together this book to help introduce you to them in a delightfully informative voice.

This volume represents a wonderful cross-section of the kinds of monsters from around the world that *might* be real. As a *Pro-Reality Activist*, I encourage you to use critical thinking skills and evidence-based reasoning to assess the plausibility of such creatures, but I'm also a huge fan of this whole field and equally encourage you to enjoy the mystery-surfing. Even after a lifetime of reading and writing about monsters I'm frequently learning about new ones and I think you'll find that Kate is a wonderful safari guide into this shadowy world of monstrous possibility.

Blake Smith – writer, researcher and podcaster

September 2021

INTRODUCTION

Bigfoot. Nessie. Mothman. The Chupacabra. Sasquatch.

Throw them all out the window. We're not talking about those celebrity cryptids. This book is about the mystery animals most people have never heard of—and there are *a lot* of them.

In early 2017 I started a weekly podcast to share my interest in animals. *Strange Animals Podcast* is about living, extinct, and imaginary animals—and I love that it's not always clear which category an animal belongs in. The first episode was about a newly reported thylacine sighting and a group of conservationists trying to breed a zebra that's as genetically close to the quagga as possible. Both the thylacine and the quagga are officially extinct... but maybe they're not, or not exactly.

This book is full of mysteries. The entries are mostly taken from the first five years of the podcast, but wherever possible I've done extra research to bring information up to date. Some of the mysteries are solved and turn out to be nothing remarkable after all, but finding a solution to a mystery is satisfying. Many more are unsolved—just waiting for more data.

A hiker might stumble across an unusual dead animal and contact an interested biologist. A grad student poking around in their university's holdings might notice a strange fossil that's clearly mislabeled. A group of scientists doing fieldwork might be about to make the discovery of the century.

However many mysteries we unravel, though, one thing is for sure: there will always be more. Every deep-sea submersible finds animals never before seen by any human. There are cave systems completely unknown to people, remote microhabitats where no human has ever set foot, key fossils brought to light after storms or floods.

Grab your hiking boots, a mosquito net, and a flashlight...or just stay in your pajamas and turn on the reading lamp. It's time to solve some mysteries!

PART ONE
MYSTERY MAMMALS

People find mammals interesting because, well, we're mammals too. People also want to pet every mammal. (Don't pet every mammal.)

Our first section covers animals like the qilin of Asia, the Nandi bear of Africa, South America's elengassen, and lots more. At least one is safe to pet.

SUCCARATH

T he Swiss naturalist and physician Conrad Gessner published the
first volume of his massive *Historia animalium* in 1551. In the book
he attempted to list every animal known, but it doesn't contain
any mention of an animal called the succarath, or su.

Gessner died in 1565. In 1603, an edition of the *Historia animalium*
published in Frankfurt had an entry for the su, a dangerous lion-like animal
with a huge bushy tail that lived in South America. The animal carried its
babies on its back and protected them with its tail, which it held over them.

Since Gessner had been dead for 38 years in 1603, he obviously didn't
write that entry himself. The publisher added it, probably because it
appeared in another popular book, this one by André Thevet.

Thevet was a French monk and scholar who lived around the same time
as Gessner. In 1555 he joined an expedition that traveled to Brazil, although
he only spent ten weeks there. In 1558 he published a book about the
strange animals, plants, and people he encountered during his travels, but
he also included information he got from sailors and other travelers. A lot of
his entries were less than accurate, but the book was popular and began
influencing other books right away. Thevet's book had an entry about the
tree sloth, for instance, and starting in 1560 reprints of Gessner's book
included an entry about the tree sloth too.

Thevet called the bushy-tailed animal the succarath and said it was found in Patagonia, which is south of Brazil. As far as anyone knows, Thevet never actually visited Patagonia. It was considered an exotic land full of wonders, including people twice the height of other humans and people with tails, so it's possible Thevet heard of the animal, didn't know where it was actually from, and assumed it was just another strange Patagonian creature.

One hint that Thevet didn't actually know where the succarath was from, and was maybe playing fast and loose with the facts, is that in his second book, he said the succarath was from Florida. That's in North America, a long way from Patagonia.

The succarath doesn't resemble any animal that lives in Patagonia. The Patagonian opossum does carry its babies on its back, but it doesn't have a bushy tail, isn't dangerous, and grows less than 10 inches long, or 25 centimeters, including its tail. It's the most southern-living marsupial in the world since it lives in southern Argentina. We don't know a lot about this opossum but it seems to mostly eat birds, mice, insects, and fruit when it can find it.

A better inspiration for the succarath is the giant anteater. It's threatened by habitat loss and hunting these days but would have been quite common when Thevet was in Brazil. The giant anteater isn't the size of a lion but it can be really big, up to 8 feet long, or almost 2.5 meters, including its tail. Its tail is long, thick, and bushy, and while it can't lift it over its back, when it sleeps it will curl up and use its tail as a blanket. It also carries its babies on its back. Specifically, it carries one baby on its back since a mother anteater only has one pup at a time. The baby's fur pattern matches the pattern of its mother's fur so that it's camouflaged while she carries it around.

The giant anteater has strong front legs with long, sharp claws that it uses to tear open ant and termite nests, and when it feels threatened it can rear up on its hind legs and slash with its claws. But the giant anteater also has a distinctive long, thin muzzle. The succarath was supposed to be a carnivore with sharp teeth while the giant anteater doesn't have any teeth at all. It licks up ants and termites with its sticky tongue and swallows them whole. Its stomach has rough folds to crush the insects, and it also swallows grit and tiny pebbles to help with the crushing process.

Another possibility is that the succarath was inspired by the southern river otter, which is endemic to Patagonia and grows almost 4 feet long including the tail, or 110 centimeters. That isn't anywhere near lion-sized but it does have a long, thick tail and will sometimes carry its babies on its back while swimming. It can be aggressive if it feels threatened. It's endangered these days due to habitat loss, pollution, hunting, and other factors, but was once relatively common. The succarath was supposed to be hunted for its fur, which is also true of the river otter, and the succarath was supposed to live near water too.

The succarath might even have been inspired by old knowledge of the ground sloth. The ground sloth Mylodon lived in much of Patagonia until about 10,000 years ago. It was actually much larger than a lion, up to 13 feet long, or 4 meters, including its thick tail. That's not the biggest ground sloth that ever lived, but it's certainly big.

It wasn't a carnivore, though. In fact, it mostly ate grass. We know for certain that's what it ate because scientists have analyzed Mylodon dung found in caves, dung that's so well preserved in the cold, dry conditions that people thought it was fresh instead of more than 10,000 years old. We also have lots of bones, fur, and even partially mummified Mylodon bodies. As a result, we know a lot about what it looked like and how it probably behaved.

Mylodon was big, heavy, and probably pretty slow-moving. Like other ground sloths, though, it had a fearsome weapon: the long, strong claws on its front feet. They were so long it walked on its knuckles. When a sloth felt threatened, it could rear on its hind legs, with its thick tail helping to prop it up, to use its claws to slash at predators. This even assumes a predator would bother attacking it, since its skin was studded with bony osteoderms that gave it built-in armor.

Mylodon and other ground sloths were related to modern tree sloths but were also related to modern anteaters like the giant anteater. Baby Mylodons might have ridden on their mother's back at least part of the time. Mylodon had a thick, hairy tail, although it couldn't lift its tail over its back, and it was big and would have been a dangerous animal to hunt. But the succarath was supposed to be a swift-moving animal, definitely not true of Mylodon.

It's quite likely that the succarath is the result of Thevet misunderstanding accounts of various animals that lived in South America. If one sailor told him about the giant anteater carrying its babies on its back, and another sailor told him the same thing about the southern river otter, he might have assumed both sailors were talking about the same animal and just combined the details. The result is an animal that never really existed, even though it appeared in just about every book about animals published in the 17th century.

～

The Lion's Tail

THE LION IS the only felid with a tuft of hair at the end of its tail, called a tassel, but the hair hides something interesting. The last bones of the tail are fused together to form a tiny spike, sometimes only a few millimeters long but sometimes almost 2 inches long, or 5 centimeters. The spike sticks out of the skin like a tiny claw. It's sometimes called a tail thorn or a caudal claw, since caudal means tail.

No one knows what the spike is for, if anything. Not every lion has one, although when it's present, both the male and female can have one. It also appears to be rather loosely attached in some lions so can be lost at some point. Cubs are born without one but it forms when the cub is around five or six months old.

The tail thorn is sometimes slightly spiky or ridged and sometimes smooth. There are occasional reports of a tail thorn in leopards and tigers, which are closely related to the lion—and even rarely in domestic cats, which are not.

Since the tail thorn is hidden by the tail tuft, it can't have anything to do with offense or defense or impressing other lions. Scientists have no idea what it's for.

ELENGASSEN

Patagonia is a big area at the southern end of South America, partly Chile and partly Argentina, with the Andes Mountains running down the western side. It's home to a number of peoples who all have legends of strange animals and animal-like beings. Modern knowledge of these beings is often poor even among the indigenous peoples after centuries of invasion and colonization. This includes a creature called the elengassen.

The first mention of the elengassen in writing came in 1866, when a Swiss naturalist named Jorge Claraz explored parts of northern Argentina. He kept a diary where he wrote about the people he met and the things he saw. The pronunciation of elengassen isn't clear since every time he wrote the word, he spelled it differently.

According to Claraz, the elengassen was supposed to live in a particular cave near the Negro River. Claraz writes that it was "[a]n animal similar to a man—it has a human figure—but is very big. It has hands, big legs, it walks like a man and is covered like a peludo [an armadillo] with an enormous hard shell, which is of stone, but the rest of the body is soft. These beings existed before, but now they are extinct. They were harmless and never attacked. But when one came near them—especially at dusk—they threw stones. These strange beings lived in caves."[1]

The cave Claraz was told about had fallen in before he visited it so he couldn't tell if there were any remains inside. A road had once passed below the cave but it was abandoned by the local people because the elengassen would throw stones and shout insults at anyone on the road. Instead, the people made a new road that detoured 3 miles around the cave, or 5 kilometers.

But there's more to the story than Claraz knew. Elengassen is an old Northern Tehuelche word that means "a young bird." Historians suggest it originally meant something like "the sun's chick" and specifically referred to a traditional hero named Elal. Elal was the son of a monster and a cloud goddess, and he taught people to use fire, hunt, and cook. He was also said to kill monsters, including eventually his own father. But as the centuries passed and the Tehuelche and other native peoples encountered each other through trade or war, Elal came to be regarded as a more monstrous entity himself, a sort of devil that people started to fear.

Elal was also described as a hunter of enormous animals as well as monsters. His tent was said to be covered with trophies, including the shells of giant armadillos. It's possible that this detail confused Claraz and he wrote that the elengassen's body had an armadillo-like shell, not his tent. Then again, other monsters in various Patagonian legends are described as having armadillo-like shells.

In the 1890s a man named Ramón Lista collected Aonikenk legends, including one about a monster who steals children and kills hunters. Elal discovered that his arrows bounced off the monster's shell. Another of the collected legends mentions a monster called the Oókempam, which "walked on four legs and was covered by a thick and very hard carapace, which was not pierced by arrows or the sharp claws of a puma."[2] Only its ankles were vulnerable.

So what exactly is going on here? Was the elengassen a real animal, a creature of folklore, or a god?

Around the end of the 19th century, when Claraz and other explorers were interviewing locals and recording folktales of the indigenous people, the cultures of the region were being subjected to a terrible ongoing shift. European invaders had killed or enslaved many of them or were in the process of doing so. Christian missionaries had also done their best to undermine the old religions. This is why there was such a disconnect between, for instance, Elal the hunter of monsters and benefactor of humankind, and the elengassen as a monstrous devil-like creature that people feared. The old stories had been forgotten or changed to reflect the people's new situation. Since the elengassen was supposed to live in caves, any cave was dangerous.

But remember also that Claraz was Swiss. He either spoke enough of the local languages to understand the gist of any conversation or had a translator, but he would probably have had trouble with some concepts, especially if the person telling him wasn't completely clear on the details to start with. Claraz's original description of the elengassen seems like a mishmash of a folklore monster and a real animal.

Claraz thought the elengassen was a glyptodon, a huge relative of the armadillo that lived in what is now Argentina and a few surrounding areas. Some species grew to the size of a compact car. Its shell was domed and covered most of its body except for its feet, tail, and head, sort of like a big tortoise shell made of interlocking osteoderms. Its tail was also armored and some species had a bony knob on the end of the tail that was probably used as a defensive weapon. Even the head and hind legs of some species were armored with osteoderms. Claraz was convinced that the glyptodon was a meat-eater, although we know now that it was a herbivore that grazed on tough plants.

Glyptodon probably had poor eyesight and may have spent the day in burrows where it was safe from predators. It went extinct around 11,000 years ago but archaeological evidence points to humans living in Patagonia before then, so humans may have hunted it for hundreds or even thousands of years. Even after the glyptodon went extinct, the enormous carapaces of dead animals remained for a long time. They were big enough that humans used them as shelters. Occasionally a carapace is discovered even today when earth is excavated by builders or paleontologists.

Much of what Claraz reports about the elengassen applies to the glyptodon: that it's covered with an enormous hard shell, that it's extinct, that it was harmless. But the glyptodon wasn't throwing rocks at anyone or standing on its hind legs like a person.

My guess is that old memories of the glyptodon persisted in local cultures for a long time but that it became confused with other animals and monsters over the centuries. Memories of the glyptodon might have vanished completely, but the armadillo remained to help remind people that giant armadillos used to exist.

It's also possible that Claraz added some details to fit his theory that the elengassen was a meat-eating glyptodon. It's too bad that we can't just ask members of the native populations that remain today, but unfortunately the ones that survived the colonization of their land lost almost all of their cultural heritage. All we have left are the writings of early naturalists like Claraz who described the native people's cultures before they were completely destroyed.

KHTING VOAR

I n 1994, a German zoologist named Wolfgang Peter visited Vietnam in South Asia. He was poking around in a market in Ho Chi Minh City when he noticed a strange pair of horns. They were black and widely spread, spiral in shape and about 18 inches long, or 46 centimeters. He didn't recognize what animal the horns belonged to, so he took pictures that he showed to his colleagues when he got home. Nobody could identify them.

Naturally, Dr. Peter grabbed another zoologist, Alfred Feiler, and they rushed back to Vietnam to look for this mysterious animal. They didn't find the animal, but they did find eight pairs of the horns. This time they bought them. Later that same year, they described the animal formally as a new species in its own genus, *Pseudonovibos spiralis*. It means spiral-horned false kouprey.

A kouprey is a wild ox from Cambodia, which has similar horns. But the kouprey's horns are oval in shape instead of round like the new bovid's horns, and the kouprey's horns are smooth instead of ridged with rings called annulations. The kouprey also hasn't been sighted since the 1970s and is probably extinct.

While Peter and Feiler were still trying to figure out what animal their mystery horns belonged to, a Norwegian zoologist named Maurizio Dioli visited northeastern Cambodia and found two pairs of spiral horns at a market. He thought they looked like the horns of a young female kouprey, but not quite. The horns he found were attached to a piece of skull at their base, called a frontlet. The skull of any animal is made up of separate pieces of bone that fit together like a jigsaw puzzle. When an animal is young the skull bones aren't tightly knit together, but as it ages the bones grow together securely, making a pattern of growth between the pieces that are called sutures. The sutures on the frontlets Dr. Dioli found were fully fused. He didn't have horns from a young female kouprey, he had horns from a fully grown animal—and he didn't know what kind.

Dioli learned that the locals in Cambodia called the animal the khting voar, which means "wild cow with vine-like horns." They described it as a slender, swift, deer-like animal that stood up to 4 feet high at the shoulder, or 1.2 meters. Some reports said that it was covered with spots. It lived in family groups in forested mountains and would sometimes stand on its hind legs to reach leaves.

The khting voar was also supposed to eat snakes. The story was that when a khting voar grabbed a snake to eat, the snake would bite the animal's horns and inject venom, which caused the ridges and twisting of the horns. As a result, people used the khting voar horns as a medicine for snakebite, and also used them in religious rituals that had to do with snake spirits and snake deities.

It didn't take too long for Dr. Dioli to learn that his khting voar horns were from the same animal Drs Peter and Feiler had discovered in Vietnam, which is right next to Cambodia. But no scientist had ever seen a live animal, or a dead animal, or the hide of a dead animal, or the full skeleton. Just the frontlets and horns.

Illustration from a 1607 Chinese encyclopedia

There were hints that knowledge of the animal went back centuries. A Chinese encyclopedia dated to 1607 shows a goat-like animal with horns like the khting voar and also mentions its snake-eating habits. Some reports from big game hunters dating back to the 1880s mention a strange spiral-horned bovid, and researchers found horns from animals killed in the 1920s.

There were also some intriguing hints that the khting voar had possibly been alive until recently. An Australian zoologist named Jack Giles, who went to Vietnam to look for the animal, was shown an old photo by a local. It showed a hunter sitting on the body of a dead khting voar, but the quality of the photo was so poor that he couldn't make out details—and the hunter was actually sitting on its head anyway, hiding any potentially useful details that might have helped Giles identify it. In December of 1994 some hunters caught a young bovid in central Vietnam, but it died before scientists could learn more about it than that it existed. By the time anyone came to study it, it had been eaten. There's no way of knowing if that animal was a khting voar or something else.

In 1999 researchers extracted DNA from horn fragments and determined that the khting voar was more closely related to goats than to antelopes or cattle. Other genetic testing indicated it was more closely related to Asiatic and African buffalo. But in 2001 a team of French biologists published papers asserting that Pseudonovibos horns were forgeries.

The French team tested DNA from six frontlets and discovered they matched those of Vietnamese domestic cattle. The horns' keratin sheaths showed manipulation to create ridges and a spiral shape. The researchers pointed out that the genetic testing that indicated the khting voar was closely related to goats had already been dismissed as being contaminated by DNA from a chamois, or rather from a piece of leather from a goat-like animal called the chamois. Chamois leather is often used for polishing things like fancy cars or trophy horns.

Bovids, which include cattle, bison, antelopes, goats, gazelles, and many other animals, grow true horns. The core of a horn is bone and a keratin sheath grows over the horn. You may have heard of drinking horns and powder horns, and those are the keratin sheaths that have been taken off the horn core of a dead animal to be used as a container.

One report says that hoaxers in Vietnam would remove the keratin sheaths off the horn cores of a domestic bull, soak them in vinegar, and then wrap them in leaves from bamboo and sugar cane and heat them up. At that point the keratin was softened enough to allow the hoaxers to twist the horns into a spiral shape and carve ridges into them. Then they'd jam the horn sheaths back over the horn cores and sell them to trophy hunters.

This all sounds like the khting voar was never a real animal, just a hoax. But is it really that simple? Of course not!

During the 1990s, a number of animals previously unknown to science were discovered in Vietnam and other parts of South Asia. This included the saola and three species of muntjac. The saola is a bovid and muntjacs are deer, so it's not out of the question that another unknown bovid was hiding in the remote forests of the area.

Not only that, but the horns of the khting voar were supposed to be of medicinal and religious value. Many animals have been driven to extinction because humans decided a part of its body was medicine. The more people will pay for an animal horn—for instance, a rhinoceros horn—the more likely that all the animals that grow those horns will be killed. When the animal is extinct, someone will figure out that you can fake those horns and sell them as the real thing.

The horns tested by the French biologists were indeed from cattle, probably produced not for sale as trophies but for sale as medicine. But I couldn't find anything that indicates the French team has genetically tested all the khting voar horns and frontlets found.

It seems likely that the fake horns were based on those from a real animal, one whose association with snakes went back for centuries. It's probably extinct now, unfortunately, both from overhunting and from habitat loss.

If it *was* a real animal, eventually more remains will turn up. Then researchers can conduct more genetic testing and determine what exactly the animal was.

If we're really lucky, a living population will turn up too.

TANKONGH

The Republic of Guinea in West Africa borders the ocean and is shaped sort of like a croissant. The middle of the country is mountainous, which is where an animal called the tankongh is supposedly found.

The tankongh is supposed to look like a small, shy zebra with tusks that lives in high mountain forests. That description may make you think of a chevrotain. The chevrotain is a small ruminant with short tusks or fangs instead of antlers. Many have white stripes and spots, including the water chevrotain.

The water chevrotain is the largest of the known chevrotain species, but that's not saying much because they're all pretty small. The female is a little larger than the male, but it's barely more than a foot tall at the shoulder, or 35 centimeters. The coat is reddish-brown with horizontal white stripes on the sides and white spots on the back. It has a rounded rump with a short tail that's white underneath. It's basically sort of rabbit-like in shape, but with long slender legs and tiny cloven hooves. It lives in tropical lowland forests of Africa, always near water. It's nocturnal and mostly eats fruit, although it will also eat insects and crabs.

The water chevrotain only lives in lowlands, though, while the tankongh is supposed to live in the mountains. But the water chevrotain is the only species of chevrotain that lives in Africa. All the others are native to Asia.

It's possible there's another chevrotain species hiding in the mountains of Guinea and nearby countries. One visitor to Guinea reported being shown some tiny gray hooves and pieces of black and cream skin supposedly from a tankongh that had been killed and eaten. Since the water chevrotain is red-brown and white, the skin must be from a different animal.

Hopefully, if this is a species of chevrotain that's new to science, it's safe in its mountain habitat from the deforestation, mining, and other issues threatening many animals in Guinea.

SOUTH AMERICAN UNICORN

I n 1977, a British writer named Bruce Chatwin published a book called *In Patagonia*. It was a travel journal from his trip to Patagonia in 1974, but parts of it appear to be heavily fictionalized. In other words, he probably made a lot of it up.

One thing he didn't make up was the theory of an animal called the South American unicorn. In his book, Chatwin writes about meeting an elderly priest named Manuel Palacios, who was convinced that the unicorn not only once lived in Patagonia, it was depicted in rock art by its ancient human hunters. Chatwin said he climbed the Andes to look for the rock art but when he found it, the animals depicted were just bulls.

The real-life priest was named Manuel Jesús Molina, who was from Argentina and who died in 1979. Molina held a number of unorthodox beliefs about the local history, including animals. He was convinced that the unicorn really did live in Patagonia, that it really was depicted in rock art, but that it went extinct thousands of years ago.

Molina published a book about the prehistory of Patagonia where he discussed some of his theories, including the unicorn. He cited two pieces of rock art and we know where both are. One is dated to 3,850 years ago and the other to at least 9,000 years ago and possibly 13,500 years. The younger piece is quite weathered and it's not clear if it represents a large-bodied animal with a long, thin horn growing from its forehead or a deer-like animal with a slender neck whose head has weathered away. The other painting is of a round-bodied animal with a short tail, slender legs, a slender neck, and small head with what looks like a single straight horn growing from its forehead.

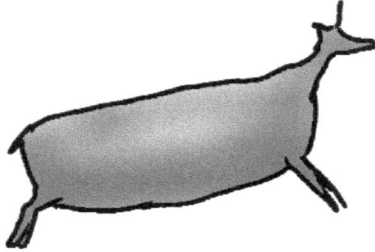

From a carving at Cueva de las Manos

Molina wasn't just going by rock art, though. He thought the rock art depicted an extinct animal called Toxodon, which really did live in South America until around 10,000 years ago. Toxodon remains have been discovered with spear tips so they definitely got killed and eaten by humans, and human predation may have played a part in their extinction. Toxodon belonged to the family Toxodontidae, in the order Notoungulata, which were hoofed mammals endemic to the Americas. That order is completely extinct now.

Toxodon probably looked something like a furry, hornless rhino but with a more cow-like head. It was a bulky animal that stood about 5 feet tall at the shoulder, or 1.5 meters, although its body was nearly 9 feet long, or 2.7 meters. The fossils we have show a heavy skeleton that would have supported a muscular body, and it walked on the flats of feet that probably somewhat resembled elephant feet. It was a herbivore that didn't specialize in any particular type of plant like grass but would happily eat anything leafy. It may have had a prehensile upper lip too. In the past, researchers thought the Toxodon was a hippo-like animal that lived in rivers and other waterways, but it actually preferred semi-arid steppes and plains.

Toxodon had no horn although Molina thought it did. A closely related animal called Trigodon, though, definitely had a short horn on its forehead. We don't know a whole lot about Trigodon except that it may have lived in swampland and preferred to eat softer vegetation. It had longer legs than Toxodon but was about the same size otherwise. Another Toxodon relative, Adinotherium, which I'm happy to report means "not terrible beast," also had a small horn on its forehead—really small in its case, not much more than a bump. It was about half the size of Toxodon and was probably a fast runner.

Trigodon went extinct seven million years ago, Adinotherium around eleven million years ago. It doesn't appear that any member of Notoungulata living at the same time as humans had a horn at all.

Most likely the rock art depicts the local deer in profile so that only one antler shows. However, there is a true unicorn that lives in South America... but it's not a hoofed animal. It's a spider.

That's right. There's a genus of spider that's actually called Unicorn that lives in high elevations in Argentina, Chile, and Bolivia. It gets its name from a pointed projection on the male's head. It's a very small spider, though, with a body only 3 millimeters long at most and long, slender legs. You're probably never going to see one. Sorry.

QILIN

T he qilin, also called the kirin, gilen, or some other close variation, is sometimes called the Chinese unicorn. These days it's usually depicted with a pair of antlers like a deer, but in older legends and artwork it often only had one horn. It can resemble a dragon with cloven hooves, or a bull-like or deer-like animal with scales or a scaly pattern on its body. Sometimes it resembles a horse or a tiger, but it almost always has cloven hooves.

The qilin legend is thousands of years old, with the first references dating back to the 5th century BCE. It has traditionally been considered a gentle animal whose appearance foretold the birth or death of a great ruler, or if it appeared to a ruler, it foretold a long, peaceful reign.

One story says it first appeared to the Emperor Fu Hsi 5,000 years ago as he walked along the banks of the Yellow River. A single-horned animal emerged from the water and walked so daintily that its cloven hooves didn't leave prints in the mud. A scroll on its back was miraculously not wet, and when Fu Hsi unrolled the scroll he saw a map of his kingdom and written characters that taught him written language.

In 1414, explorer Zheng He brought a giraffe to China for the first time and presented it to the emperor as a qilin. The emperor wasn't fooled, but it was a good PR move to treat the animal as a qilin. But the qilin was never depicted with a long neck before then, and even after, long-necked qilins were rare in art and sculpture. On the other hand, the word for giraffe in Japanese and Korean is kirin.

The qilin was supposed to be solitary and lived high in the mountains and in deep forests, where it ate plants. It's long been linked with the unicorn, which was probably inspired by garbled accounts of rhinoceroses. The qilin may also be inspired by the rhino or even the giraffe, or both, but there's another animal that might have contributed to its legend.

Tsaidamotherium was a bovid that lived during the late Miocene, as much as seven million years ago or possibly as little as half a million years ago. Its fossils have been found in northwestern China. It was probably most closely related to the musk ox and was adapted for life in cold mountainous regions. It had a high nasal cavity, which would have helped warm air before it reached the lungs. Other bovids found in cold areas tend to have similar structures. It was probably a browser, eating leaves, seeds, and other plant material.

The saiga and the takin have a pair of clearly separated horns. The saiga's horns are long and look like typical antelope horns, while the takin's horns resemble those of a musk ox, curving to the sides in a sort of U shape.

The most striking thing about Tsaidamotherium is its horns, and no animal living today has horns even slightly like it. It had a pair, but only the right horn grew large. The left one was much smaller, so that from a distance it looked like it only had one large horn on its head. These are not slender unicorn horns, though. They're not even bull-like horns.

There are actually two species of Tsaidamotherium that we know of and they had differently shaped horns. *T. hedini* had thick horns that grew upward from the head like cones. The other, *T. brevirostrum*, with fossilized remains only described in 2013, had the same mismatched horns but both were short and squat and bent forward something like a Phrygian cap. Both males and females had the horns. The two known species lived in different parts of the mountains and may have been much more widespread than is currently known.

T. brevirostrum (left) and T. hedini

There are hints that Tsaidamotherium may have survived into the modern era, possibly in isolated pockets in the Himalayan Mountains. Until the mid-19th century there were reports of animals matching *T. hedini*'s description in Tibet, although even there it was considered rare to the point of near-legend. The first fossilized remains of Tsaidamotherium weren't discovered until the early 20th century, in 1932.

One interesting note is that the larger horn of *T. hedini* would probably have resembled the conical Yeti skullcaps sometimes found in monasteries in the Himalayas. But that's probably a coincidence.

WAMPUS CAT

The wampus cat, or just wampus, has appeared in folklore throughout North America for over a hundred years and probably much longer, especially in mountainous areas in the eastern portion of the continent.

The term actually comes from the word catawampus, probably related to the phrase catty corner or kitty corner. Both words mean "something that's askew or turned diagonally," but catawampus was also once used in the southeastern United States to describe any strange creature lurking in the forest. It was a short step from catawampus to wampus cat, possibly also influenced by the word catamount, used for the cougar and other large cats native to North America.

Whatever the origins of the word, the wampus cat was usually considered a real animal. Some people probably used the term as a synonym for catamount, but many people firmly believed the wampus was a different animal from the cougar, bobcat, or lynx. It was usually supposed to be a type of big cat, although not necessarily.

The word wampus also once referred to a dress-like garment resembling a knee-length smock worn over leggings, also called a wampus coat. The first newspaper use of wampus referring to an animal doesn't appear until the very end of the 19th century. A Missouri paper wrote in May 1899, "They knew immediately the source of the hair-raising scream. The 'wampus' was after them. They could see it; it was a big black thing with long hair and large feet."

What may be a follow-up to that story, from a different Missouri newspaper, appeared in November 1899 and was headlined "THE WAMPUS IS DEAD."

Many described it as gray wolf, but others refused to believe such an animal was here and lightly spoke of the wampus. It frequented the dark woods at day time, coming forth at night and roaming around, uttering a strange cry. Woe unto the traveler overtaken by darkness, for the night was made hideous by the shrill cry. [...]

On last Sunday night George Jolliff secreted himself in a tree south of his house and about 7 o'clock saw the long-sought monster, accompanied by several dogs, approaching; on seeing his dog, which was tied beneath the tree, they come under the tree and Mr. Jolliff fired, severely wounding the animal. Hastily climbing down, he fired again and this stopped the monster in his tracks. [...]

It measured 6 feet 7 inches in length and 2 feet 4 inches in height; some say it is a female wolf, others a cross between the dog and wolf species. It is dark brown tinged with red and black.

This sounds like a coyote or red wolf, especially considering that it was accompanied by dogs. The size of the animal in metric is 2 meters long, presumably including the tail, and 73 centimeters tall. The height is accurate for a coyote or red wolf but it's much longer than even a gray wolf. It's possible the animal had an unusually long tail or there was a little exaggeration going on. Whatever the animal was, dogs, coyotes, and wolves don't make a screaming sound, so this wasn't the same animal that kept frightening people with its shrill cries.

By the beginning of the 1900s, the wampus as an animal had completely overtaken the use of the word as an article of clothing. The first baseball team named the Wampus Cats, from Texas, appears in 1908, which argues that the term wampus cat had been in common use for some time. The surge of articles about the wampus cat also suggests the term had made its way from local use into the popular culture. By the 1910s, any unidentified animal is referred to as a wampus, from a striped rodent-like animal to an exhibit in a traveling menagerie, which unfortunately wasn't described. Humorous articles claiming to answer the question of "what is the wampus?" appear alongside humorous poetry about the wampus. Well, it supposed to be humorous but it's not actually funny.

By the 1920s, newspaper reports of the wampus cat were routine. Its description varies and most reports are light on definite details. Here are some examples:

- November 1897 (near Clarksville, Tennessee): "Mr. Gaisser was within ten feet of the strange animal and describes it as being about six feet in length, of a ferocious appearance, having long claws and looking as though it could attack and dispatch a man as easily as a hog. [...] What kind of an animal it is Mr. Gaisser cannot say. It has the appearance of being either a jaguar, mountain lion or a catamount."
- November 1918 (near Vestal, Tennessee, a community in south Knoxville): "It looked very much like a leopard. It was a short haired animal, with a slick, glossy coat. It was white and gray spotted, and had a long tail, with a bushy end."
- December 1921 (in Howell County, Missouri): "Drake says it was a long lanky animal, had spots on it. Then Bill Webb saw the 'wampus cat.' It was in the day time. Bill says it was running and disappeared in a second. It was built like a tiger and light yellow in color, he reports."
- January 1926 (near the Spring Creek community of Crenshaw County, Alabama): "The animal has been seen by a number of people, and apparently either is a panther or a monster wildcat from the recesses of Patsalega swamp. The beast is described as being of the size of a large shepherd dog, dark of hue and shaggy of coat. It steps from eighteen to twenty inches while walking, and when running it covers the ground in huge leaps of from six to ten feet. It has long claws, and leaves a footprint measuring two and one-half inches in width." (That's about 6.5

centimeters, which is not a very big pawprint for a big cat; a cougar's print would be up to 4 inches across, or 10 centimeters.)

By the 1940s, newspaper mentions of the wampus as an animal diminish, taken over by sports teams with the name. By the 1960s wampus cat articles are mostly space-filling pieces talking about traditions among local old-timers, usually with a humorous tone although again, they're not funny, and fewer sports teams carry the name. By the 2000s, when reporters were doing their research online, any mention of a wampus cat is accompanied by the (bogus) Cherokee story about a woman who could turn into a wildcat, and usually also claim that "wampus" is a Cherokee word.

Even though the term wampus fell out of favor slowly after its peak in the late 1920s, reports of cat-like animals still appeared in newspapers. They just didn't get called wampus cats. Here are a few from the 1950s and 60s.

- March 1950 (Nebraska): "I wasn't more than 20 rods [110 yards, or 100 meters] from him. He was reddish in color, about three feet high, and had a bushy mane (small) behind his head. His head was wide and he had broad shoulders."
- January 1951 (Pennsylvania): "Sent to hunt a strange animal reported sighted in the Noxen-Harveys Lake area, three bloodhounds today were themselves the objects of an extensive search in that region. Meanwhile, other reports of 'strange' animals came from the Hazleton area. The three bloodhounds... began trailing the animal, variously described as a bobcat, lynx, or mountain lion...."
- February 1955 (Pickens, South Carolina): "A resident...reported today he had seen the animal yesterday and described it as being black and having long hair and a long bushy tail. Mr. Findley said he heard weird sounds about 10 o'clock last night and went out with a light and gun, but neither saw nor heard anything more. He said the sounds were similar to a huge cat."
- July 1962 (Cherrytree Township, Pennsylvania): "According to Mr. Black, the large animal jumped from the limb of an oak tree on his farm and fled into the cover of a nearby game preserve. [...] He pointed out the marks on the ground where the animal

landed and inspected claw marks on the tree. [...] Asked to compare its size with that of a large dog, the farmer said it was considerably larger and tawney coated."

- February 1963 (near Roan Mountain in the Cherokee National Forest, south of Elizabethton, Tennessee): "It has a track larger than a big dog, and is black in color. A real shiny black. Most of the dogs refuse to track or bay this strange animal and those with nerve enough to get close to the animal wish they hadn't. Mr. Birchfield had one dog that was real brave and ventured close, but the poor dog was carried home by Birchfield with broken bones."
- September 1965 (Fairview, North Carolina, between Thomasville and Lexington): "The 'animal,' described as dark in [color], resembling a cat but much larger, was first seen when the Thomasville-Lexington reservoir was under construction in the late 1950's. [...] It has been reported that the cries sound 'like cats fighting, then ending with the sound of a bob-white bird.'"

Reports still happen today, posted online. In a May 2018 comment on an article about wampus cats, someone named Greg Brashear writes, "I saw what the old farmers in my area in north central Ky. [Kentucky] call a 'wompus cat'. [...] It was bigger than a bobcat but smaller than a cougar with yellow eyes and a [disproportionately] long tail and it was solid black."

The cougar (also called a mountain lion, puma, painter, catamount, or panther) was once common throughout most of the Americas but was hunted to extinction in much of the eastern United States around the early to mid-20th century. It's a big animal, able to kill deer, with a big male weighing as much as 100 kilograms, or 220 pounds. It can leap enormous distances—up to 40 feet while running, or 12 meters, up to 18 feet straight up into a tree, or 5.5 meters—and can sprint up to 50 miles per hour, or 80 kilometers per hour. It doesn't roar but instead produces an unearthly scream. It can also purr.

There's no doubt that at least some wampus cat reports were cougars. Cougar sightings have continued in the eastern United States and Canada through the present day. Young male cougars travel widely to establish a territory, so most modern sightings in the southeast are probably of young males who have traveled from populations in the west. There's evidence that many more cougars have started moving into the northeastern United States and Canada and may have even established breeding populations.

The cougar varies in color from tawny to reddish and is sometimes grayish-white. Occasionally a white or partially white individual is caught on camera traps, but there has never been a confirmed sighting of a black cougar. It's likely that sightings of wampus cats described as yellow or gray and white are actually cougar sightings.

Another North American cat, the jaguar, is sometimes black in color. Jaguars are fairly common in South America but much less common in Central and North America. North American jaguars are also much smaller than South American populations. The jaguar strongly resembles the leopard, with rosette-like black spots on a tawny or yellowish background. Melanistic jaguars are usually called black panthers and are rare in the North American population.

By 1960 the jaguar was almost completely extirpated in the United States, but a population remains in northern Mexico and occasionally one roams across the border into Arizona, Texas, or New Mexico. It prefers heavily forested areas near water sources.

It's possible, although very unlikely, that a young male black panther could roam as far as the eastern United States and contribute to wampus cat sightings, although the jaguar doesn't fit wampus cat descriptions very well. The jaguar roars instead of screaming like the cougar, and its extreme shyness makes it unlikely to venture close to humans.

There is a third possibility, assuming the wampus cat isn't an animal new to science. The jaguarundi is also native to the Americas, including most of South and Central America through northern Mexico. It's related to the cougar and is solid colored, without spots, with a coat that can be black, gray-brown, or reddish. It's only about twice the size of a domestic cat but looks much different.

The jaguarundi's body is long and its legs are relatively short in proportion, which means it has a somewhat otter-like gait when it runs. Its rounded head has small round ears, also resembling an otter's. Its tail is long, thick, and bushy. It lives in forests, rainforests, open areas (as long as there's brush to hide in), deserts, and mountains across a wide range, but it's not very well studied.

The jaguarundi used to be found in the United States, although its former range is unclear. Confirmed and/or credible sightings have been reported in Texas, Arizona, Alabama, and especially Florida, including roadkill animals. It's mostly nocturnal but is somewhat active during the day as well, and has at least 13 different calls, including whistles, growls, screams, and chattering and chirping.

This is interesting when compared to some of the wampus cat sightings. Brashear's 2018 comment about a black cat with a disproportionately long tail sounds like a potential jaguarundi, although he described it as bigger than a bobcat when the bobcat is typically larger (although not as long, especially if you include the tail). The 1965 report of a dark-colored cat whose screams ended with a bird-like call (the bobwhite makes a two-tone whistle) also sounds like the jaguarundi.

It's exciting to think that many wampus cat sightings might be of the jaguarundi, especially since sightings continue to the present day. Fortunately, the jaguarundi is a protected species in the United States and throughout most of its range. It would be great if these interesting wild cats were found to have established breeding populations in the less populated areas of the United States.

WATER ELEPHANT

I n 1912, an article appeared in the *Journal of the East Africa and Uganda Natural History Society*. It was written by someone named R.J. Cuninghame but concerned a Mr. Le Petit.

A lot of stories about strange African animals from this era are hoaxes of one variety or another, and this one is doubly suspect because it's a second-hand account. The article claimed that Le Petit had seen a so-called water elephant twice, both times in what is now the Democratic Republic of the Congo and the Republic of the Congo in central Africa.

The first time he saw it was around June 1907 when traveling down the River Congo in a canoe. He saw the head and neck of a large animal surface out of the water, then sink again soon after. The native people with him said it was a water elephant.

The second time, he saw five of the animals out of the water, although the article doesn't say when this was. He described them this way:

Height at shoulder, 6 to 8 feet [1.8 to 2.4 meters]. Legs relatively short. Back curved, as in *E. africanus*. Tail not observed. Neck *about* twice the length of *E. africanus* with ears similar in shape to those of that species, but relatively smaller. Head most distinctly long and ovoid in form, with trunk only about 2 feet [61 centimeters] in length. The shape of the feet was seen in the spoor on sand and showed four toes distinctly separated as in the hippopotamus, but the weight of the body seemed to be carried by the toes largely, while the plantar impression of the sole was not very pronounced. The ground was level where this spoor was seen. All the animals observed had no traces of tusks. Skin is apparently hairless, smooth and shiny resembling that of a hippopotamus, only darker. The gait was elephantine, and the last seen of those five water-elephants was their disappearance into the water, which was deep. In habit they are nocturnal, coming out to feed on strong rank grass after sundown. They spend the day in the water much as hippopotami do.

This 1912 article isn't actually the first account of a water elephant, and not even the first account of Le Petit's sightings, although it's the most detailed. Rumors of an African species of tapir had been around since at least 1898 and probably quite a bit longer, circulating among British big game hunters.

The problem is that the tapir is definitely not an African animal. There are four known species of tapir alive today, three in Central and South America and one in Asia.

While the different species vary in size and coloring, generally the tapir has a rounded body with a pronounced rump, a stubby little tail, and a long head with a short but prehensile trunk. Superficially it looks kind of like a pig but it's actually much more closely related to horses and rhinos. It has four toes on its front legs, three on its hind legs, and each toe has a large nail that looks like a little hoof.

The tapir is a shy, largely solitary, mostly nocturnal animal that prefers forests near rivers or streams. Its favorite method of hiding is to submerge in water. It spends a lot of time in water, in fact, eating water plants and cooling off when it's hot. It swims well and can use its short trunk as a snorkel. Technically

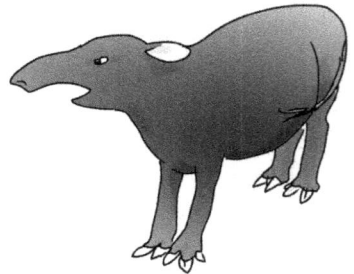

the trunk is called a proboscis and the tapir mostly uses it to help gather plants. Its ears are relatively small and rounded.

While the tapir does superficially fit the description of the water elephant, it's nowhere near a perfect match. The Asian tapir is the largest species alive today and it only stands about 3 feet 7 inches tall at the very most, or 110 centimeters. Le Petit claimed his water elephants were twice that height.

The tapir evolved in North America and later spread into South America and Eurasia. Tapir fossils have been discovered in Europe, China, and the Americas, but not Africa. That doesn't necessarily mean tapirs have never lived in Africa, just that we haven't found any fossils.

We do have fossils of an interesting animal called Moeritherium. It lived about 35 million years ago and its fossils have been found in northern and western Africa. It was related to modern elephants although it wasn't a direct ancestor, just an offshoot that as far as we know died out without descendants.

Like the tapir, Moeritherium looked more like a pig than an elephant. It stood between 2 and 3 feet high at the shoulder, or up to about 92 centimeters tall, but was long-bodied, almost 10 feet long, or 3 meters. Its legs were short, it may have had a tapir-like trunk, and it had small tusks similar to those of a hippo. Studies of its teeth indicate it ate a lot of aquatic plants, so it probably lived a lot like a hippo.

Could the water elephant be a descendant of Moeritherium? It sounds like a possibility at first, but if Moeritherium survived to the present day, why did the hippo become so widespread to fill the same ecological niche? The hippo evolved about 16 million years ago. Sixteen million years is a long, long time for two animals to compete for the same resources in the same areas.

There is another possibility. Le Petit might have seen the African forest elephant, sometimes called a pygmy elephant since it's smaller than the African bush elephant. It stands about 8 feet tall, or 2.4 meters. It also lives in the Congolese rainforests and, like other elephants, spends a lot of time in the water and can swim well.

Then again, Le Petit specified that the water elephant only had a trunk about 2 feet long, or 61 centimeters. If his report is accurate instead of a hoax or a mistake, he must have seen an animal that's completely unknown to science.

JAVA ELEPHANT

Borneo is the largest island in Asia and the third-largest island in the world. It's home to hundreds of organisms found nowhere else in the world, if not thousands, although many are recently extinct or endangered due to habitat loss from logging and palm oil plantations. The island still supports a few thousand elephants, though—but no one's sure how the elephants got to the island.

In 1750 or thereabouts, according to locals, a pair of elephants was given to the Sultan of Sulu, who brought them to Borneo. At some point the elephants were released into the wild and their descendants now live throughout the western and northern parts of the island.

This story sounds straightforward and interesting, but there are a lot of confusing details that make it less certain. Supposedly, the Raja of Java gave a pair of elephants to Raja Baginda of Sulu, but that was around 1395. We do know that in 1521, tame elephants were part of the palace's wonders, but by the 1770s there were no tame elephants, only wild ones. Supposedly, the elephants were released into the wild at some point to keep them from being captured in the event of an invasion.

Whenever and however it happened, it sounds plausible that the elephants still living in Borneo are descendants of elephants gifted to a local ruler. Elephants have long been considered appropriate royal gifts. The story is given more weight by the fact that no elephant fossils have ever been found in Borneo, which suggests the elephants were introduced recently. The Bornean elephants also have very low genetic diversity, which would be the case if they were descendants of a single pair.

But here's why these smallish, rather tame elephants in Borneo are such a big deal. Locals, and some researchers, think they're the only surviving members of a subspecies of Asian elephant, called the Java elephant, that went extinct by the 1800s.

The Borneo elephants are different in appearance and behavior from other Asian elephant subspecies. They're slightly smaller, although they're not actually pygmy elephants as they're sometimes called. A big male Borneo elephant may stand about 8 feet tall at the shoulder, or 2.4 meters, while a big male Asian elephant may reach close to 10 feet, or 3 meters. The Borneo elephant's tusks are straighter than other Asian elephants—some males don't have tusks at all—and their tails are so long that in some individuals, they actually touch the ground.

Borneo and Java are part of the Malay Archipelago in Southeast Asia. Java is over 800 miles south of Borneo, or almost 1,300 kilometers, so it's not like the elephants could get there in modern times without human help.

In 2003, DNA testing on the Borneo elephants indicated they were not related to other subspecies of Asian elephant and were either from Java or native to Borneo. It also suggested that the elephants had been on Borneo for a really long time, possibly as much as 300,000 years.

A more recent genetic analysis, published in *Nature* in January 2018, concludes that the Bornean elephant is indeed a separate subspecies of Asian elephant. It doesn't speculate how long elephants have lived on Borneo, but it does show a genetic bottleneck in the population that dates to between 11,000 and 18,000 years ago.

This makes sense from what we know of the region. Borneo was cut off from the Asian mainland and the rest of the Malay Archipelago around 18,000 years ago, when sea levels rose due to melting glaciers. There might not have been very many elephants in the area when Borneo was finally separated from the rest of the world.

The report also points out that Borneo is a tropical area and conditions aren't favorable for fossils to form. Elephant fossils might exist and just haven't been found yet, too. But the paper brings up another mystery that it can't solve. Elephants only live in western and northern Borneo, not throughout the whole island and not near the point where Borneo was once connected with other islands when sea levels were lower.

We still don't know whether the Borneo elephant is the same as the extinct Java elephant. No one has been able to sequence the Java elephant's DNA profile for comparison.

~

Kallana

THE INDIAN ELEPHANT is a subspecies of Asian elephant that lives throughout much of mainland Asia. It's smaller than the African elephant but still pretty big, with males standing as much as 11.3 feet at the shoulder, or 3.4 meters, although most are much smaller than this. Females are smaller than males and have smaller tusks, or sometimes no tusks. It was once common throughout India but is now endangered due to habitat loss and poaching.

Reports of a suspected dwarf elephant species called the kallana come from southern India. The kallana elephant reportedly only grows to around 5 feet high, or 1.5 meters, and while it looks like an ordinary Indian elephant except for its size, it doesn't mix with Indian elephants and even appears to avoid them. It lives in rocky hills in and around the Peppara Wildlife Sanctuary in southern Kerala. It's shy and can move much faster than regular elephants, and it doesn't appear to have trouble with steep slopes the way elephants usually do.

In 2005, a wildlife photographer named Sali Palode got pictures of two kallana elephants, one alive, the other a dead one he found by a lake. He took more photos in 2010, and in 2013 he got brief video footage. But there are no photos of a herd of kallana elephants, just solitary animals. Without being able to examine a kallana elephant in person, researchers don't know if the elephant photographed is a new species or subspecies, or just an Indian elephant with a genetic anomaly similar to dwarfism in humans. The photos might even just be of young elephants that haven't grown to their full size yet.

A granite statue of an elephant that stands about 3 feet high, or about 92 centimeters, is located near the entrance to the Padmanabha Swamy temple in Kerala. It was recently restored and raised to street level, since for a long time it was mostly buried. It's at least 200 years old and may be closer to 2,000 years old. There's always a possibility that the presence of a small elephant statue at such an important temple influenced folktales of the kallana elephants. Then again, it's also possible that the kallana elephants inspired the carving of a small statue. Or, of course, the two might be completely unrelated.

Until someone gets definitive footage of a herd of kallana elephants, an individual is captured and studied, or someone takes samples of the elephant dung found throughout the hills and sends it for genetic testing, there's no way of knowing if the reported small elephants are something special.

LINNAEUS'S ELEPHANT

I n 1753 the Swedish scientist Linnaeus got to examine a fetal elephant preserved in a jar of alcohol. Back then hardly anyone outside of Asia and Africa had seen an elephant, so Linnaeus was enormously excited about it and wrote to a friend that the specimen was as rare as a diamond.

Linnaeus described the species and named it *Elephas maximus*, also known as the Asian elephant today. But from records that still survive, the specimen was marked as having come from Africa. A Dutch pharmacist and collector had acquired the specimen around 1736 and after he died it was sold to King Adolf Frederick of Sweden, who let Linnaeus examine it. The auction catalog where it was listed for sale indicates that it was from Africa, but in his official description of the elephant Linnaeus wrote that it was from Ceylon, now called Sri Lanka, which is in Asia.

Ever since there's been a mystery as to whether the elephant specimen was actually an Asian elephant or an African elephant, and if Linnaeus even knew that there were elephants in Africa. Because the specimen is of a fetal elephant—that is, a baby that died before it was fully developed, probably when its mother was killed while she was pregnant—it's hard to tell just by looking if the specimen is an African or Asian elephant. We do still have the specimen, fortunately, which is held in the Swedish Natural History Museum's collection.

A mammal expert at the London Natural History Museum, Anthea Gentry, got curious about the specimen in 1999 when she saw it on a trip to Sweden. Gentry's husband was a paleontologist who specialized in mammals, and later she showed him a photograph of the specimen and asked what he thought. He said he was pretty sure it was an African elephant, not an Asian elephant. Gentry got permission to do DNA testing on the specimen, but since it had been in alcohol for so long, not even the most advanced technology and the world's most experienced expert in ancient DNA could get a usable genetic sequence from the tissue.

The world's most experienced expert in ancient DNA was Tom Gilbert of the University of Copenhagen in Denmark. He did his best and failed, but he couldn't forget about the little mystery elephant. In 2009 he got an idea for extracting genetic material from the specimen in a new way that might yield results. It took years, but he and his team got it to work. In 2012 the mystery was finally solved. Linnaeus's little elephant was actually an African elephant.

But that's not the end of the story. When a scientist describes a new species and gives it its scientific name, the first specimen described is known as the type specimen. Linnaeus's elephant was the type specimen of the Asian elephant—but once it was proven to be an African elephant, it couldn't *continue* to be the type specimen of the Asian elephant. But that meant the Asian elephant had no official type specimen.

When an animal is described officially, it's a formal process. The International Commission on Zoological Nomenclature decides whether a suggested name is acceptable and makes decisions on type specimens and taxonomy. Researchers connected with the Commission started digging around for a new type specimen, preferably one from Linnaeus's time or earlier.

A type specimen isn't always a whole animal. A lot of times it's just a little piece of a skeleton or a partial fossil, although the more complete a specimen is, the better. Linnaeus had described a partial elephant tooth at some point which was still available in a Swedish museum, and taxonomists were considering using that as a type specimen when they got an email from a paleontologist who specialized in elephants. He sent a copy of a travel journal from an amateur naturalist named John Ray, who had visited Florence in 1664 and wrote his observations of an elephant skeleton and skin on display in the duke's palace.

And, it turned out, the elephant skeleton John Ray had described was in the collection of a museum in Florence. And it was definitely the skeleton of an Asian elephant—in fact, we even have what amounts to a photograph of the elephant when it was alive, because none other than the artist Rembrandt sketched it. So that skeleton was designated as the type specimen of the Asian elephant and all is well.

NANDI BEAR

Sightings of the Nandi bear have come from various parts of Africa but especially from Kenya, where it's frequently called the chemosit. There are lots of stories about what it looks like and how it acts. Generally, it's supposed to be a ferocious nocturnal animal that sometimes attacks humans on moonless nights. Some stories say it eats the person's brain and leaves the rest of the body. That's creepy. Also, just going to point this out, it's extremely unlikely.

The Nandi bear's shaggy coat is supposed to be dark brown, reddish, or black, and sometimes it will stand on its hind legs. When it's standing on all four legs, it's between 3 and 6 feet tall, or about 90 centimeters to 2 meters. Sometimes it's described as looking like a hyena, sometimes like a baboon, sometimes like a bear. Its front legs are often described as powerful.

The first known sighting by someone who actually wrote down their account is from the *Journal of the East Africa and Uganda Natural History Society*, published in 1912. The account is by Geoffrey Williams.

The Nandi expedition Williams mentions took place in 1905 and 1906. It was a military action by the British colonial rulers who killed over 1,100 members of the Nandi tribe in East Africa after the tribe basically said, hey, stop taking our land and resources and people. During the campaign, livestock belonging to the Nandi were killed or stolen, villages and food stores burned, and the people who weren't killed were forced to live on reservations. Here's what Geoffrey Williams had to say about the Nandi bear, which suddenly doesn't seem quite so important than it did before I learned all that.

Several years ago I was travelling with a cousin on the Uasingishu just after the Nandi expedition... [J]ust as we drew near to the slopes of the hill, the mist cleared away suddenly and my cousin called out 'What is that?' Looking in the direction to which he pointed I saw a large animal sitting up on its haunches not more than 30 yards away [27.5 meters]. Its attitude was just that of a bear at the 'Zoo' asking for buns, and I should say it must have been nearly 5 feet high [1.5 meters]. It is extremely hard to estimate height in a case of this kind; but it seemed to both of us that it was very nearly, if not quite, as tall as we were. Before we had time to do anything it dropped forward and shambled away towards the Sirgoit with what my cousin always describes as a sort of sideways canter. The grass had all been burnt off some weeks earlier and so the animal was clearly visible.
I snatched my rifle and took a snapshot at it as it was disappearing among the rocks, and, though I missed it, it stopped and turned its head round to look at us. It is in this position that I see it most clearly in my mind's eye. In size it was, I should say, larger than the bear that lives in the pit at the 'Zoo' and it was quite as heavily built. The fore quarters were very thickly furred, as were all four legs, but the hind quarters were comparatively speaking smooth or bare. This distinction was very definite indeed and was the first thing that struck us both. The head was long and pointed and exactly like that of a bear, as indeed was the whole animal. I have not a very clear recollection of the ears beyond the fact that they were small, and the tail, if any, was very small and practically unnoticeable. The colour was dark and left us both with the impression that it was more or less of a brindle...

A couple of years later, in the same journal, a man saddled with the name Blayney Percival wrote about the Nandi bear. "The stories vary to a very large extent, but the following points seem to agree. The animal is of fairly large size, it stands on its hind legs at times, is nocturnal, very fierce, kills man or animals." Percival thought the differing stories referred to different animals, known or unknown.

One night in June of 1960, a man named Angus Hutton was driving home to his tea estate in Kenya and saw two Nandi bears in the headlights as he rounded a bend. At first he thought they were actually small bears standing on their hind legs, about 4 or 5 feet tall, or 1.2 to 1.5 meters. Then he realized both had all four feet on the ground but were quite tall animals.

Hutton shot the larger one, although the smaller ran off into the forest. The dead animal was gingery-brown in color with long, shaggy hair, had small ears and a short, thin tail, and its shoulders were much higher than its hindquarters. Hutton also noted some confusion about the animal's sex. Its penis was large but the animal also appeared to be a female.

Hutton took pictures of the animal and prepared the body for study and, ultimately, taxidermy. Then he shipped the cleaned skeleton, skin, and plaster casts of the footprints to a friend at the Coryndon Museum in Kenya. The friend responded with a letter thanking Hutton and stating the animal was the long-haired brown hyena.

The brown hyena is rare and lives only in southern Africa, at least as far as we know. Aside from that, though, the brown hyena doesn't really match up with the animal Hutton shot. It's smaller and has larger pointed ears, a bushy tail, and is dark brown with long fur. Hutton's Nandi bear had small rounded ears, a tail that was only tufted at the tip, and a reddish coat.

There's another animal that does match Hutton's Nandi bear description, though: the spotted hyena.

The spotted hyena is the largest hyena species, generally up to 3 feet tall at the shoulder, or 92 centimeters. Females are larger than males. The spotted hyena is indeed spotted, although the color and pattern of its coat varies in different individuals. It has small rounded ears, a tail that's tufted at its tip, and it can be reddish-brown in color, although yellowish or pale brown is more common.

But the reason why Hutton's Nandi bear matches so closely with the spotted hyena is its genitals. Spotted hyenas live in matriarchal clans, meaning that females are leaders of the group and are more socially important than males, and the female spotted hyena's genitals look like a male's. She has what's called a pseudo-penis and a pseudo-scrotum, the only species of hyena that does. That would explain why Hutton was confused as to the animal's sex when examining it. Not only that, the spotted hyena does live in Kenya, unlike the brown hyena.

All that remains is for scientists to take another look at Hutton's Nandi bear, maybe even retrieve a genetic sample from it and compare it to the spotted hyena, right?

That would be the case, but the specimen can't be found. We're pretty sure it was received at the museum, but no one knows where it is. The photos Hutton took are missing too. Most likely, they're all in storage and haven't been looked at in over fifty years. Hopefully someone will come across them while looking for something else, and any lingering mystery of the Nandi bear's identity can be solved.

GIANT BEARS

The Kamchatka Peninsula is in the very eastern part of Russia on the Pacific coast. It's mountainous with a cluster of active volcanos and is well known for the brown bears that live in the area.

The Kamchatka bears are among the largest brown bear subspecies in the world, almost the size of the closely related Kodiak brown bear. When one stands on its hind legs it can be almost 10 feet tall, or 3 meters. The Kamchatka brown bears also have long brown fur.

But in 1920 a Swedish scientist named Sten Bergman was shown a bear pelt by locals that was jet black and had short fur. Not only that, it was much larger than a brown bear pelt. Bergman also saw a huge skull supposedly from one of the black bears and a paw print 15 inches long and 10 inches wide, or 38 centimeters by 25.5 centimeters.

No one has reported any giant black bears in the area since, living or dead. It's possible that the bear was an unusually large brown bear with anomalous fur. Brown bears do have considerable variability in both the color and length of their fur.

The Kamchatka Peninsula has another mystery bear too. In 1987 a hunter named Rodion Sivolobov bought a giant white bear skin from locals. It looked like a big polar bear pelt, but the locals assured him it was from a very specific, very rare type of local bear.

They called it the "pants pulled down" bear because of its short hind legs, which made it look like it was wearing furry trousers that were sagging low around its knees. It was supposedly large but with a relatively small head, with fur that was white or grayish. Because of its short hind legs, it had an unusual method of running. It supposedly ran in a sort of bounding motion more like a rabbit than a typical bear, and was supposed to be slow as a result.

Sivolobov sent samples of the pelt to various zoologists in Russia, but they said there wasn't much they could determine without a skull. Later he also got hold of a skull that was supposed to be from one of these rare white bears, but examination by experts indicated that it looked like an ordinary brown bear skull.

It's not clear if Sivolobov still has the pelt and skull these days and if not, what happened to them. People have suggested that the white bear might just be a stray polar bear that traveled to the Kamchatka Peninsula on ice, which has been well documented as happening occasionally. It might also be the offspring of a polar bear and a brown bear, also not a rare occurrence. Or, more excitingly, it might even be a surviving descendant of the Eurasian cave bear that supposedly only went extinct about 24,000 years ago—although that's much less likely.

Whatever it is, there are no modern reports of the "pants pulled down" bear.

WOLF MYSTERIES

Andean Wolf

In 1927, a German animal collector called Lorenz Hagenbeck bought a wolf pelt in Buenos Aires, Argentina. The seller said the pelt, and three others, came from a wolf-like wild dog in the Andes Mountains.

The pelt is about 6 feet long, or 1.8 meters, including the tail, with thick, long fur, especially a thick ruff on the neck. It's black on the back and dark brown elsewhere.

Hagenbeck didn't recognize the pelt, so when he got home he sent it for examination. In the 1930s and 1940s, various studies suggested it belonged to a new species of canid, possibly one related to the maned wolf. One mammologist, Ingo Krumbiegel, also thought he might have seen a skull of the same canid in 1935, which he said had resembled a maned wolf skull but was much larger, and which was supposed to have come from the Andes. Krumbiegel was convinced enough that in 1949 he described the Andean wolf formally as a new species. But no more specimens have come to light.

In 1954 another study determined Hagenbeck's pelt was just a dog pelt, possibly of a German shepherd crossbreed. A 1957 study came to the same conclusion. In 2000, a DNA analysis came back inconclusive due to the pelt having been chemically treated during preparation and due to contamination with dog, wolf, human, and pig DNA. Currently the pelt is on display at the Zoological State Museum in Munich.

~

Honshu wolf

THE JAPANESE WOLF, or Honshu wolf, is a canid that's supposed to have gone extinct in January of 1905 when the last known individual was killed. But people keep seeing and hearing it in the mountains of Japan.

The Honshu wolf was a small subspecies of gray wolf, not much more than a foot tall at the shoulder, or about 30 centimeters. Its legs were short and its short coat was grayish-brown. It was once considered a friend to farmers since it ate rats and other pests. Wolves were also regarded as protective of travelers in Japanese folklore. But in 1732 rabies was introduced to Japan. That disease, combined with loss of habitat, made the Honshu wolf more of a threat to humans and their livestock, which led to its persecution.

Sightings of the wolf have continued ever since the last one was killed in 1905. Photographs of a canid killed in 1910 were studied by a team of researchers in 2000, who determined that the animal in the photos was probably a Honshu wolf. People have found tracks, heard howling, seen wolf-like animals, even taken photos of what look like wolves. The problem is that the Japanese wolf looked similar in many ways to some Japanese dog breeds like the Shiba Inu and the Akita. People might be seeing dogs roaming the countryside.

We can't even DNA test hairs and old pelts to see if they're from wolves, because we don't have a genetic profile of the Honshu wolf. There are only a few taxidermied specimens of the wolf and none of them have yielded intact DNA.

~

Ringdocus

IN 1886 A MONTANA settler named Israel Hutchins shot a wolf-like animal that had reportedly been killing livestock. No one knew what it was, so Hutchins traded it to a taxidermist for a cow. He needed the cow because when he first tried to shoot the wolf creature, he accidentally shot one of his own cows instead.

The taxidermist, Joseph Sherwood, also owned a general store in Idaho. He displayed the stuffed canid in the store, where it stayed for almost a hundred years until it disappeared. In 2007 Hutchins's grandson, Jack Kirby, traced it to the Idaho Museum of Natural History.

The stuffed mystery canid is usually called the ringdocus, a name Sherwood made up. It has a sloping back and some other unwolf-like features that might be due to bad taxidermy or might be due to physical anomalies in an ordinary wolf or dog—or might be due to the ringdocus being an animal new to science. Suggestions as to its identity include a thylacine, a hyena, a wolf-coyote hybrid, a wolf-dog hybrid, or a dire wolf.

It's not a thylacine, I'm just going to say that straight out.

Since we have the taxidermied specimen, it seems logical that a DNA test would clear up the mystery or bring us a brand new scientific mystery, if it turns out to be an unknown animal. But Kirby doesn't want a DNA test done. That tells me it's probably just a wolf, and he knows it's a wolf.

In May 2018, also in Montana, a rancher shot a wolf-like animal and couldn't figure out what it actually was. The animal had long gray-brown fur, a large head and huge ears, but legs that appeared shorter than an ordinary wolf's legs. The rancher decided to get a DNA test done and the results indicated it was an ordinary gray wolf. Wolves, dogs, and other canids can show great physical variation from individual to individual.

That doesn't mean there aren't any unknown canids in the world, just that DNA testing is important to figure out what's a mystery and what's a wolf.

GOLDEN RAT

In December 2004, a woman named Karen Stoker was walking her dogs near Kenwood Pond in London's Hampstead Heath when she saw something unusual in the water. She said it was swimming in a straight line fast enough to leave a wake. She described it as orange and bigger than a rat, but smaller than a fox. She also said it didn't have a tail, or at least she didn't see one. When the animal saw her, it hurried out of the water and disappeared into tall grass.

The story appeared in the local news and people immediately started making suggestions as to what the animal might be. Let's look at a few of these suggestions and evaluate them against the little we know about the sighting, and see if we can figure out the animal's identity.

One suggestion is that it might be a ferret. A feral domesticated ferret or a wild ferret, also known as a polecat, can swim, but it's not a water animal. A ferret out hunting would stay in the grass, not get in the water, and it probably wouldn't get in the water at all in

winter. Ferrets also aren't orange, although a reddish coat crops up very rarely in polecats. But the long body shape of a polecat, ferret, or related animal would be instantly recognizable. Stoker didn't mention that the animal was long and skinny like a ferret or weasel.

Another suggestion is that it was a European water vole, an uncommon animal in Britain although it's common in other parts of Europe and Asia. The water vole lives in burrows it digs in the banks of ponds and slow-moving streams and rivers. It eats plants although it will occasionally eat frogs and tadpoles too. In years where there's a lot to eat, so many water voles are born that they can destroy all the grass and other plants available around their waterways.

The water vole is a rodent that sort of resembles a chubby rat with a shorter nose and a tail that's covered with fur. But it's smaller than a rat on average, only about 9 inches long at most, or 22 centimeters, not counting its tail, which is about half the body's length. The brown rat is about 11 inches long, or 28 centimeters, not counting the tail. Stoker described the animal she saw as bigger than a rat but smaller than a fox. The water vole is also chestnut brown, not orange.

What animals are orange and of the right size and might be swimming in a pond in England in December? Stoker would have recognized a fox, even one without a tail. Even if an exotic animal like a red panda escaped from its owner and got lost, it wouldn't have been swimming around in December. Besides, red pandas have big bushy tails and are big chonks that would look larger than a fox. A cat can be orange but cats aren't known for their interest in swimming, and of course Stoker would definitely recognize a cat.

In the news account, she described the animal this way: "My goodness, it's orange!" In other words, it was probably pretty bright in color. But we don't know the time of day when this sighting took place or what the weather was like. On some winter evenings, the sunset light seems to make everything glow. A russet brown animal might appear orange.

Maybe it was an animal that's not native to the UK, a muskrat. The muskrat is a rodent native to North America but introduced to Europe and Asia as a fur animal. It's not that much bigger than a rat but is much heavier, so might appear larger. It does have a long, thick tail that's flattened on the sides to help it swim, but the tail is black, and instead of fur it's covered in scaly skin. When it gets out of the water, the tail drags on the ground so it might be easy to miss.

The muskrat is semi-aquatic and well adapted to the water, where it spends most of its time. Its fur is short but dense to keep it warm even in cold water. It can seal its ears closed when it dives and it can stay underwater for over fifteen minutes at a time. It has webbed hind toes with stiff bristles around the edges, although it also uses its tail to help it swim. It can even swim backwards.

While it's most active at dawn and dusk, the muskrat feeds at any time of the day or night and is active all year round. It constructs trails through underwater plants which it can follow even when the water freezes. It often shares its habitat with the much larger beaver, which it somewhat resembles, but it's actually more closely related to the hamster than to the beaver. It's also closely related to voles and in fact is considered a type of vole.

But the muskrat isn't orange. It's generally brown or black, although it can be a reddish-brown color. While there were once populations of muskrats in the UK, introduced for their fur, they were trapped and killed off in the 1930s. There are plenty of muskrats still living in Europe, though, so many that they're an invasive pest in some areas.

It's possible that a stray muskrat somehow ended up in London and lived in the large Kenwood Pond briefly, until it was eaten by a predator such as a hawk or fox. As far as I know there was only the one sighting, which argues that if this was an unusual animal like a muskrat, it didn't live long. Either the animal was a reddish color that appeared orange in the sunlight, or it might have been a rare blond animal with its wet fur appearing orangey. While blond muskrats are rare in the wild, it's possible that a fur farm somewhere has been breeding for the color.

There is another possibility. The coypu, or nutria, is another semi-aquatic rodent that has been imported to Europe for its fur, although it's native to South America. It looks like a muskrat but is bigger, up to 2 feet long, or 61 centimeters, not counting its tail. It mostly eats plants and is most active in the twilight. While it usually lives around slow-moving streams, shallow lakes, and in wetlands like the muskrat, the coypu will also tolerate saltwater wetlands. Wild coypus are generally dark brown, but ones bred for their fur are often blond or even white.

The coypu has been sighted in the UK much more recently than the muskrat. Attempts to eradicate escaped coypu in Great Britain were ongoing from the 1940s through the 1980s, until 1989 when only three males were trapped. That seemed to be the last of the coypu in England, but in 2010 one was spotted in Ireland. In 2015 some pet coypus escaped and have formed a breeding colony in Ireland which continues to spread, despite attempts to trap them all.

The coypu is larger than a rat and a large individual could be as big as a fox, but a more average one would be somewhere between the two animals. Ones bred for the fur trade or as pets often have light-colored fur that could easily appear orange in the right light when it's wet. It swims well and fast enough to leave a wake, and while it does have a tail, it wouldn't be noticeable in the water and is short in comparison to its body. It's also much less tolerant of cold weather than the muskrat, so if one did escape a fur farm or a pet owner released it in December, it might easily die of cold before it was sighted by very many people.

One interesting thing about the coypu is that the female has teats that are high up on her sides, which allows her babies to nurse even when they're all in the water. Also, speaking of bright orange, the coypu's big incisor teeth are bright orange because of the iron in its teeth enamel that makes the teeth strong.

VESPUCCI'S GIANT RAT

T he Florentine explorer Amerigo Vespucci reached Brazil in 1503, and while he was there he visited the volcanic island Fernando de Noronha. He wrote about his visit later and one of the things he mentioned was that the island was home to very large rats.

Since Vespucci was the first European ever to visit the island, and no one from anywhere in the world was living on it at the time, the rats he saw can't have been the rats he was used to. That would have been the black rat, since the brown rat hadn't spread throughout Europe yet. It did so later, outcompeting the black rat in most environments.

Other explorers and sailors visited the island in the years after 1503, and by 1888 when biologists came looking for the very big rat, all they found were the descendants of black rats brought there by ships.

Then, in 1973 paleontologists from Brazil and the United States visited the island to see what had once lived there—and they found remains of the very large rat.

It turns out that the rat wasn't actually a rat, although it was a rodent, and while it was larger and heavier than the black rat, it wasn't enormous. It was about the size of a typical brown rat, in fact.

Vespucci's rat has been placed in its own genus and named *Noronhomys vespuccii*. Reseachers think its ancestor might have been semiaquatic like some of its rodent relatives that still live in South America. Rodents who spent a lot of time in the water would have occasionally been swept out to sea and floated or swam to the island. Once a population of the rodents was established on the island, they evolved to be exclusively terrestrial.

Ironically, Vespucci's rat was probably driven to extinction by the ship rats that colonized the island soon after Vespucci visited.

~

Rat Kings

A RAT KING isn't one animal but a group of rats joined together by their tails. This sounds like something out of folklore but it's actually a real occurrence, although it's rare. The oldest report known dates to 1564, but specimens are occasionally uncovered even today. All reliable reports of rat kings are of black rats. The black rat has a long, thin, flexible tail that it uses to help it climb.

Not much is known about how rat kings form, but the most widely accepted suggestion is that a group of rats huddling together for warmth get their tails tangled together without realizing it. When each rat tries to separate itself from the group by pulling, the knot tightens. Eventually the rats are permanently stuck together.

It seems reasonable to think that a bunch of rats stuck together by their tails wouldn't survive long. They'd starve to death or kill each other trying to get free. But a rat king made up of seven rats found in the Netherlands in 1963 was examined and even X-rayed to learn more about it, and where the tails were intertwined there was evidence of calluses forming. This suggests the rats may have survived for some time.

Most rat kings are made up of young rats, possibly siblings sharing a nest. It's probable the mother of the 1963 rat king fed them and kept them alive until they were discovered by a farmer, who killed them.

Rats aren't the only animals found with their tails knotted together. It happens to squirrels occasionally too. In the case of squirrels, pine sap and nesting material can glue or tangle the tails of young squirrels together. We have video evidence from 2013 and 2018 of modern squirrel kings, along with the evidence of veterinarians who managed to separate the squirrels in both cases so they wouldn't die.

HORNED HARES AND JACKALOPES

When someone prepares a dead animal for taxidermy, it's not a simple process. The taxidermist has to remove the skin from the body, clean it and add preservatives, make a careful armature or mannequin of the body from wood or other materials, and put the skin on the armature and sew it up. The taxidermist then adds details like glass eyes and artificial tongues. It can take months of painstaking work to finish a specimen, and it requires a lot of artistry and training.

Taxidermists who are learning the trade will often mount small, common animals like rabbits and rats as practice. Sometimes they'll get creative with the process, just to make it more interesting. If a taxidermist adds pronghorn horns to a jackrabbit, voila, it's a jackalope!

You can see stuffed jackalopes today in a lot of places, since they're fun conversation pieces. Some restaurants will have one stuck up on a wall somewhere, for instance. Horned hares are similar, but instead of a jackrabbit with pronghorn horns or white-tailed deer antlers, which are animals from North America, the European horned hare is usually a European hare with horns from a roe deer.

The horned hare was once such a common taxidermied animal that people actually believed it was real. Around the 19th century, as knowledge of the natural world grew more sophisticated, scientists realized rabbits and hares don't have horns and those stuffed specimens were just hoaxes. The tip-off was probably when taxidermists started getting really fancy by adding bird wings and saber teeth to their mounted hares.

But...

The horned hare goes way back in history. It appeared in medieval bestiaries, sometimes called the unicorn hare. The unicorn hare was supposed to have a single black horn on its head. The hare would act normal but when someone approached, it would spring at them and stab them with its horn. Then it would eat them.

The legend of the horned hare is so widespread and long-lived, in fact, and was believed for so long, that maybe it was based on something real.

There is a strange truth behind jackalopes and horned hares. A disease called the Shope papilloma virus, or SPV, affects hares and rabbits. There are a lot of papilloma viruses in various animals, but in most animals, including humans, it only results in tumors inside the body. In rabbits and hares, it causes keratinized tumors to grow from the skin, often on the head. Usually these are small and don't show through the fur, but sometimes an animal has an extreme case of SPV and it genuinely looks like it has horns. The horns are hard and usually dark in color. As if that wasn't bad enough, rabbits and hares in Europe can also get a disease called Leporipoxvirus that again causes facial horns to grow from the skin.

If you're feeling totally creeped out right now, don't worry. Humans can't catch these diseases from rabbits and hares. Also, rabbits and hares, even ones with "horns," don't kill and eat people.

WINGED CATS

Winged cats are a real phenomenon but the wings in question are furry, not feathered, and winged cats can't fly. That doesn't stop people from claiming they've seen these winged cats flying. In Ontario, Canada in 1966 a so-called vampire cat was supposedly flying around attacking other animals. It was a black tomcat with furry wings 7 inches long, or 18 centimeters. Eventually someone shot the cat, which was examined by a veterinarian and found to be rabid. Its wings were nothing but thickly matted fur so the stories of it flying around weren't true.

In 1959, a case went to court in West Virginia over a winged cat. A 15 year old boy named Douglas Shelton said he'd rescued the cat from a tree and adopted her, but a woman named Mrs. Hicks said that the cat was hers, named Mitzi. Mitzi had run away and Mrs. Hicks wanted her back. At first the judge awarded the cat to Mrs. Hicks, but when Douglas brought her into the courtroom, she had no wings. Douglas said she'd shed them during the summer but he'd kept the wings, which he showed to the judge. At that point, Mrs. Hicks suddenly decided she didn't want the cat after all. Frankly, I'm sure Mitzi was better

off with Douglas, who didn't care if she had wings or not, although he did change Mitzi's name to Thomas.

Stories like these didn't just happen back in the olden days. There are lots of winged cat reports today, including photos and videos. Why do some cats develop these furry appendages that people call wings?

Sometimes the cats in question just have long fur that becomes badly matted and appears to form winglike flaps along the sides. In other cases, the wings are due to a rare skin condition called feline cutaneous asthenia, or FCA.

Cats with FCA have unusually elastic skin. All skin stretches at least a little bit but almost immediately snaps back into place. You can try this yourself by gently tugging up the skin on the back of your hand and releasing it. In cats with FCA, the skin doesn't snap back properly, especially the skin along the shoulders and back. Since in the ordinary course of living its life, a cat's skin stretches quite a bit along the back, eventually an FCA cat ends up with long flaps of furry skin that stretched out and just stayed that way. The wings aren't really wings, of course, and can't allow the cat to fly.

Cats with FCA usually need special care, especially if the case is severe. The skin is prone to damage because it's actually very delicate. The so-called wings sometimes tear off naturally, leaving wounds that bleed very little but still need to be treated by a veterinarian. The wings then reform. They tend to be on the sides near the hind legs but are sometimes closer to the shoulders.

Mitzi, AKA Thomas, was definitely a cat with FCA. Her wings were 6 inches long, or 15 centimeters, and her tail was said to be squirrel-like. She was a white cat described as a Persian, although she may have just had long hair like a Persian cat. A reporter who examined Thomas described her wings as fluffy at the ends but with a gristly feel at the base, as though they contained tendons or other structures. This was probably the extended skin due to FCA.

It sounds like Douglas was a nice kid who rescued a cat from a tree and took her home. When his friends made fun of the unusual-looking cat, he was really upset.

Once word of the winged cat got around, people started showing up at the family's house to look at it. At first Douglas charged ten cents to see the cat, and he was even invited to New York where he and Thomas appeared on the "Today Show." But after that, things started to go kind of nuts. Thousands of people kept trying to see the cat, so many that Douglas's mom spread the story that the cat had died, just so people would leave the family alone. She also took the cat to a friend's house for a while until the fuss died down. Then Mrs. Hicks sued.

Ultimately, Douglas and his family were awarded custody of Thomas by the judge, with Mrs. Hicks rewarded a single dollar in damages. I'm sure Thomas lived a good life with the Sheltons. Douglas would be about the right age to be a granddad by now, so I bet he tells his grandkids stories about the time he had a cat with wings. I bet they don't even believe him.

PART TWO
STRANGE BIRDS

Birds aren't just feathery things that go tweet. Some of them are *mysterious* feathery things that go tweet. This section includes a mystery eagle of North America, the forest raven of Europe (hint: not even slightly a raven), the pink-headed duck of Asia, and the biggest penguin *ever*.

PAINTED VULTURE

Plenty of vulture species live in the Americas, from the enormous Andes condor of South America and its cousin the California condor, to the much smaller turkey vulture that's common all the way down to the southern tip of South America. But it's possible that until only a few hundred years ago, another type of vulture lived in Florida, a species or subspecies that disappeared so completely that almost no one now believes it existed. If it weren't for two men from the 18th century, we wouldn't know about it at all.

Before we learn about that mystery vulture, though, we need to learn a little bit about a different bird, the king vulture of Central and South America.

The king vulture is a bird of lowland tropical forests, although it's most closely related to the Andean condor. It has a wingspan up to 7 feet across, or 2 meters. It's mostly white with a black or dark brown tail and flight feathers, and the ruff around its neck is black too. Its head is brightly colored in orange, deep purple, yellow, and gray; its eye is white surrounded by vivid orange; and it has a bright orange, funky-shaped wattle on its bill. It's a large, strong vulture that can open up a carcass with its heavy bill, which means other species of vulture depend on it to help them get at the good parts of a dead animal. Don't think about that too much.

The fossil record shows that other species closely related to the king vulture once lived in North America, although not for several million years. As far as we know, the king vulture has never lived in North America except for an occasional sighting in southern Mexico.

That's funny, though, because an 18th century naturalist named William Bartram described a vulture he saw in Florida. It sounds remarkably like a king vulture except that its tail was white tipped with black instead of all black. Bartram called it the painted vulture and said it was fairly common in Florida. He even shot one himself and wrote a description of it later:

> The plumage of the bird is generally white or cream colour, except the quill-feathers of the wings and two or three rows of the coverts, which are of a beautiful dark brown; the tail which is large and white is tipped with this dark brown or black; the legs and feet of a clear white; the eye is encircled with a gold coloured iris; the pupil black.[1]

William Bartram traveled through Florida in the 1770s and took notes about the animals and plants he saw and the native people he met. He published a book of his travels in 1791. In the book, he doesn't just describe the painted vulture, he devotes a few paragraphs to its habitat and habits.

Some people think Bartram got the painted vulture mixed up with a different bird, the northern caracara, a bird of prey which only looks slightly like a king vulture. Those people don't explain how he could have described a king vulture so perfectly (except for the difference in tail color) if he was actually seeing a much different bird.

Other people think Bartram saw a king vulture but misremembered the detail of its tail color. That in itself would constitute a mystery, though, since the king vulture has never lived in Florida as far as we know. Since Bartram's description was precise in its details, it's probable that he was working from careful notes of a bird he saw in person. Tail color is not something he would have left out during his observations.

The most likely scenario is that there was a population of king vultures in Florida that later went extinct, possibly a subspecies of king vulture with a mostly white tail instead of all black.

It's not just random birdwatchers and mystery animal enthusiasts who think the painted vulture was a real bird. A 2013 paper in *Zootaxa* by biologist and condor expert Noel Snyder and an expert on William Barton, Joel Fry, pointed out that "none of the arguments offered historically against the validity of the Painted Vulture is persuasive when examined closely. [There is] a strong case for acceptance of Bartram's Painted Vulture as a historic resident of northern Florida and likely other adjacent regions."[2]

As Snyder and Fry point out in the article, Bartram isn't the only person who reported seeing the painted vulture. In 1734 an English naturalist and artist, Eleazar Albin, painted a vulture that looked almost identical to the one Bartram described thirty-odd years later, tail and all. It's not known where Albin saw his bird, although Albin himself claimed he had visited Jamaica in 1701, but as far as researchers can determine Bartram wasn't aware of the painting.

So it's possible that a subspecies of king vulture once lived in Florida but went extinct sometime after Bartram saw it in the 1770s. If he and Albin hadn't documented it, no one alive today would have any idea the painted vulture ever existed.

FOREST RAVEN

The story of the forest raven starts some 470 years ago, when a scholar and physician named Conrad Gessner, who lived in Switzerland, published a book called *Historia animalium*. The book wasn't like the medieval bestiaries of previous centuries, in which fantastical and real animals were listed together and half the information consisted of local superstitions. Gessner was an early naturalist, a scientist long before the term was in general use. *Historia animalium* consisted of five volumes with a total of more than 4,500 pages, and in it Gessner attempted to describe every single animal in the world, drawing from classical sources such as Pliny the Elder and Aristotle as well as his own observations and study.

The book contained animals that had only recently been discovered by Europeans at the time, including animals from the Americas and the East Indies. It also included a few entries which no one today believes ever existed, like the fish-like sea monk and sea bishop. Those and similar monsters were probably added by Gessner's publishers against his will

or maybe just without him knowing, since he was seriously ill by the time the volume on fish was published. For the most part, though, the book was as scholarly as was possible in the mid-16th century and was lavishly illustrated too.

Volume three, about birds, was published in 1555 and included an entry for a bird Gessner called the waldrapp, or forest raven. But the illustration didn't look anything like a raven. The bird in the illustration has a relatively long neck, a crest of feathers on the back of its head, and a really long bill that ends in a little hook. Gessner wrote that the bird was found in Switzerland and was good to eat.

As an interesting side note, the illustrator seems to have been Eleazar Albin, who also illustrated the painted vulture from the previous chapter and a mystery macaw from the next chapter.

Here's a rough translation of the original Latin from Gessner's entry about the forest raven:

> The bird is generally called by our people the Waldrapp, or forest raven, because it lives in uninhabited woods where it nests in high cliffs or old ruined towers in castles. Men sometimes rob the nests by hanging from ropes. It acquires a bald head at maturity. It is the size of a hen, quite black from a distance, but if you look at it close, especially in the sun, you will consider it mixed with green. The Swiss forest raven has the body of a crane, long legs, and a thick red bill, slightly curved and 6 inches [15 centimeters] long. Its legs and feet are longer than those of a chicken. Its tail is short, it has long feathers at the back of its head, and the bill is red. The bill is suited for poking in the ground to extract worms and beetles. It flies very high and lays two or three eggs. The young ones are praised as an article of food and are considered a great delicacy, for they have lovely flesh and soft bones. Those who rob the nests of young take care to leave one chick so the parents will return the following year.

All that sounds like a perfectly ordinary bird, although not a raven. But what was it? No one knew, and eventually scholars decided that Gessner must have included a bird that didn't exist.

But it did sound like one particular bird, just not one found in Europe.

That's the northern bald ibis, which was once common across the Middle East and northern Africa.

Here's a description of the Northern bald ibis. Let's see how it matches up with Gessner's forest raven.

The Northern bald ibis is a fairly large bird, about a foot long, or 30 centimeters, with a wingspan of 4.5 feet, or 135 centimeters. That's about the size of a goose. It has black feathers that shine with iridescent colors in sunlight, including bronze, violet, and green. It has long, dull red legs and a long, curved bill that's also reddish. Its head is the same shade of dull red and has no feathers, but it does have a crest of long feathers on the back of its head and neck. It nests on cliff ledges and prefers to hunt for food in areas where the grass or other vegetation is short, such as pastures, fallow fields, semi-arid steppes, and golf courses, often 10 miles or more from the cliffs where it nests, or 15 kilometers. It eats insects and other small invertebrates, but it especially likes lizards and beetles. It probes into soft, sandy soil with its bill to find most of its food. The birds live in small flocks and often fly in a V formation.

The northern bald ibis mates for life. The male finds a good nesting site and tidies it up, then waits to see if he can attract a female. The female inspects the site and the male to decide if she likes them, and if she does, the pair build a nest of twigs lined with grass, and the female lays two to four eggs.

Oh, and the northern bald ibis is sometimes also called the waldrapp, just as Gessner reported.

All this information certainly sounds like the same bird Gessner described. But the northern bald ibis doesn't live in Switzerland or other parts of Europe. It's only known from the Middle East and northern Africa. Right?

That's what people after Gessner thought, until 1941. That's when a team of scientists excavating ancient sites in Switzerland found the bones of what turned out to be northern bald ibises. The bones were only a few hundred years old. More remains, both fossil and subfossil, have since been found in France, Germany, Austria, and Spain, and the bird probably lived in even more areas.

It turns out that the northern bald ibis was once common in many parts of Europe, especially around the Alps. It was considered a sacred bird in ancient Egypt along with the sacred ibis, and was supposed to be one of the birds released by Noah during the great flood to help him find land, so was venerated by people of different faiths in the Middle East. But in Europe, it was just considered good to eat. The Archibishop Leonard of Salzburg called for its protection in the Swiss Alps as long ago as 1504, but by the early 17th century, only a matter of decades after Gessner's book was published, the bird was extinct in Europe. It didn't take long for Europeans to forget it even existed.

Unfortunately, the northern bald ibis is still endangered due to hunting, habitat loss, and poisoning from pesticides. It's also sometimes electrocuted when it lands on electricity pylons that aren't insulated for birds, although efforts are underway to make pylons bird-safe in many areas. A successful captive breeding program has been in place since the late 1970s and that's a good thing, since the last known migratory population went extinct in 1989 and the remaining non-migratory colonies declined to only a few hundred individuals.

The breeding program has gone so well that birds started being reintroduced in some areas of their former range in about 2003, including Spain, Germany, Austria, and Italy. Tagging of the remaining wild birds has also revealed that a small population still migrates from the Middle East to Africa to winter in central Ethiopia. In some areas, conservationists have added nesting platforms to the existing cliffs so that more birds can nest safely. Hopefully their numbers will continue to climb.

I'll finish with a final piece of trivia about the northern bald ibis that I think you'll like. It's a member of the pelican family. Have a nice day.

ALBIN'S MACAW

The artist Eleazar Albin is connected with a third mystery bird, a macaw. It's sometimes called the Martinique macaw, although we don't know for sure that it was native to this island or that it even existed at all. It might even be two separate species of bird.

Macaws are a type of parrot native to the Americas. They have longer tails and larger bills than true parrots and have face patches that are mostly white or yellow. There are six living species of macaw but many others that are extinct or probably extinct. The largest living species is the hyacinth macaw, which is a beautiful blue all over except for yellow face markings. It can grow over 3 feet long, or about 92 centimeters, including its long tail. It mostly eats nuts, even coconuts and macadamia nuts that are too tough for most other animals to crack open, but it also likes fruit, seeds, and some other plant material. Like other parrots, macaws are intelligent birds that have been observed using tools. For instance, the hyacinth macaw will use pieces of sticks and other items to keep a nut from rolling away while it works on biting it open.

The story of the Martinique macaw starts almost 400 years ago, when Jacques Bouton, a French priest, visited the Caribbean in 1639 and specifically Martinique in 1642. Bouton wrote an account of the people and animals he saw, including several macaws that don't quite match any birds known today. One of these is the so-called Martinique macaw, which he said was blue and saffron in color. Saffron is a rich orangey yellow.

In the early 20th century, a zoologist named Walter Rothschild read Bouton's account and decided those birds needed to be described as new species, even though there were no type specimens and no way of knowing if the birds were actually new to science.

Rothschild was an interesting guy, to say the least. He was stupendously rich and correspondingly was eccentric in a way that ordinary people don't typically manage. He kept tame zebras that were trained to pull a carriage, for instance. He described many species of animal new to science and had many others named after him.

Life goal.

People have been trading macaws and parrots as a type of currency for thousands of years, since their large, brightly colored feathers were in high demand for ceremonial items. They're relatively easy to tame and can be kept as pets. The five birds Rothschild described from Bouton's account might have been known species that were being kept as pets outside of their natural range, hybrid individuals, or Bouton might just have made mistakes when giving details.

We do have some paintings that might be depictions of the mystery macaws. Albin painted a blue and yellow parrot with a white face patch in 1740 that's supposedly the Martinique macaw, although Albin would have seen the bird in Jamaica when he visited in 1701, not Martinique. The two islands are about 1,100 miles apart, or almost 1,800 kilometers.

A similar blue and yellow macaw appears in

Roelant Savery's 1626 painting of a dodo. The dodo lived on the island of Mauritius in the Indian Ocean, nowhere near the Americas. Savery just liked to paint dodos and included them in a lot of his art. In another 1626 painting, called "Landscape with Birds," he included a dodo, an ostrich, a chicken, a turkey, a peacock, ducks, swans, cranes of various kinds, a cassowary, and lots of other birds that don't live anywhere near each other. On the far left edge of the painting there's a blue macaw with yellow underparts.

Rothschild described the Martinique macaw in 1905 but reclassified it when he published a book named *Extinct Birds* in 1907. He got an artist to paint a depiction of it based on Bouton's account and it actually doesn't look all that similar to Albin's and Savery's birds. It's dark blue above, bright orange underneath, and only has a small white patch next to its lower mandible instead of a big white patch over the eye.

In other words, Albin's macaw might be a totally different bird from the Martinique macaw.

There is a known bird that might have inspired Albin's painting. The blue-and-gold macaw lives in many parts of northern South America. It has rich yellowy-gold underparts and is a brilliant aqua blue above. It matches the colors of Albin's painting pretty well, but not the facial markings. The blue-and-gold macaw has a white face but a large stripe of black, outlining the white patch, that extends under its chin. Albin's macaw doesn't have any black markings and its white patch is much smaller than the blue-and-gold macaw's.

Of course, Albin may have gotten details wrong in his painting. Even though he was probably painting from sketches and notes he took during his visit to Jamaica, about forty years had passed since he actually saw it. As for Savery's paintings of a similar macaw, he never traveled to the Americas and probably based his paintings on pet birds brought to Europe by sailors and missionaries. He was known for his meticulous detail when painting animals, though, and his birds clearly show the white face and black stripe under the chin of a blue-and-gold macaw, even though the blue plumage appears much darker than in living birds. This is probably due to the paint pigments fading over the centuries.

Savery's birds lack one detail that blue-and-gold macaws have: a small patch of blue under the tail. This would be an easy detail for the artist to miss, though, especially if he finished the painting's details without a real bird to look at. Albin's painting also lacks the blue patch.

That still leaves us with two bird mysteries. Was Albin's macaw a real species or just a blue-and-gold macaw with incorrect details? And what bird did Bouton see in Martinique?

If either bird existed, they seem to be extinct now—but they might not be. Many macaws live in South America in sometimes hard to explore terrain. While many known species of macaw are threatened with habitat loss and hunting for feathers or for the pet trade, there's always a possibility that an undiscovered species still thrives in remote parts of the Amazon rainforest.

WASHINGTON'S EAGLE

W e only have two known species of eagle in North America, the bald eagle and the golden eagle. Both have wingspans that can reach more than 8 feet, or 2.4 meters, and both are relatively common throughout most of North America. But we might have a third eagle, or had one only a few hundred years ago. We might even have a depiction of one by the most famous bird artist in the world, James Audubon.

In February 1814, Audubon was traveling on a boat on the upper Mississippi River when he spotted a big eagle he didn't recognize. A Canadian fur dealer who was with him said it was a rare eagle that he'd only ever seen around the Great Lakes before, called the great eagle. Audubon was no slouch as a birdwatcher and was familiar with bald eagles and golden eagles. He was convinced the "great eagle" was something else.

Audubon made four more sightings over the next few years, including at close range in Kentucky where he was able to watch a pair with a nest and two babies. Two years after that he spotted an adult eagle at a farm near Henderson, Kentucky. Some pigs had just been slaughtered and the eagle was looking for scraps. Audubon shot the bird and took it to a friend who lived nearby, an experienced hunter, and both men examined the body carefully.

According to the notes Audubon made at the time, the bird was a male with a wingspan of 10.2 feet, or just over 3 meters. Since female eagles are generally larger than males, that means this 10-foot wingspan was likely on the smaller side of average for the species. It was dark brown on its upper body, a lighter cinnamon brown underneath, and had a dark bill and yellow legs.

Audubon named the bird Washington's eagle and used the specimen as a model for a life-sized painting. Audubon was meticulous about details and size, using a double-grid method to make sure his bird paintings were exact. This was long before photography, remember.

So we have a detailed painting and first-hand notes from James Audubon himself about an eagle that…doesn't appear to exist.

Audubon painted a few birds that went extinct afterwards, including the ivory-billed woodpecker and the passenger pigeon, along with less well known birds like Bachman's warbler and the Carolina parakeet. To add to the confusion, though, Audubon also made some mistakes. Many people think Washington's eagle is another mistake and was just an immature bald eagle, which it resembles.

The largest bald eagle ever verifiably measured had a wingspan of 8 feet, or 2.4 meters, with unverified reports of 9-foot wingspans, or 2.75 meters. A bald eagle will actually have a slightly broader wingspan as a juvenile than as an adult because of the way its feathers are arranged, but that difference is a matter of a few inches, not feet. In addition, the largest bald eagles are found in Alaska. Individuals in the southeastern United States are usually much smaller. Female bald eagles are as much as 25% larger than males.

But here we have a male eagle shot in Kentucky with a measured wingspan of 10.2 feet. Juvenile bald eagles do travel widely, but even if that happened to be an outrageously large individual who'd flown down from Alaska, consider that Audubon had seen the same type of large eagle nesting a few years before near the same area. He'd watched a pair feeding two chicks.

There are plenty of differences between juvenile bald eagles and Audubon's Washington eagle beyond just size. Audubon made careful notes of the appearance of his specimen, from plumage coloration and particulars of the feet and bill, to the size and appearance of internal organs. A comparison of his notes and the known variations of a juvenile bald eagle's appearance show numerous discrepancies.

Audubon's "Bird of Washington"

Audubon kept diaries of his birding trips so we know he was familiar with juvenile bald eagles. He even painted one. We also know he differentiated between juvenile bald eagles and Washington's eagle, which he wrote was about a quarter larger than the juvenile bald eagle.

Golden eagles also resemble juvenile bald eagles to some degree, but they don't nest in Kentucky. Their winter range just barely touches Kentucky, in fact. They nest in Canada and in the western half of the United States. The largest golden eagle ever measured was a captive-bred female with a 9.3 foot wingspan, or 2.8 meters. Like bald eagles, golden eagle females tend to be considerably larger than males. Even if an aberrantly large male golden eagle decided to vacation a little farther south than usual, it's clear from many details in Audubon's painting and in his notes that the bird he examined can't be a golden eagle.

From other reports we know it hunted differently from bald eagles, including no reports of it stealing fish from ospreys the way bald eagles frequently do, because bald eagles are jerks. Washington's eagle reportedly preferred to nest in rocky cliffs near water, not in trees like bald eagles. Even the way Washington's eagle flew differed from the bald eagle.

All that aside, Audubon wasn't the only person to have reported the eagle. A Dr. Lemuel Hayward of Boston reportedly kept a live one in captivity sometime in the early 19th century, and a number of mounted specimens were displayed in museums. Audubon even examined one museum specimen and agreed that it was a Washington's eagle. It was supposedly donated to the Academy of Natural Sciences of Philadelphia at some point, but no one knows where it is now.

Audubon's juvenile "White-Headed Eagle"

Audubon's own mounted specimen has been lost too. If Washington's eagle was a real species of eagle, it's possible there are other specimens floating around in personal collections or museum storage rooms, mislabeled as juvenile bald eagles.

Then again, we have to face another aspect of the mystery. Audubon was patriotic, as evidenced by his naming the eagle after George Washington. His journals and letters are full of praise for Washington, who died in 1799, only fifteen years before Audubon first saw the "great eagle." There's always a chance that Audubon wanted to name a bird after his idol, but not just any bird. It had to be majestic and bold, the largest eagle in the world! Maybe he decided to invent one.

What most reports of the Washington eagle don't mention is that Audubon killed his specimen at a farm very close to his own house. Maybe it did happen that way, or maybe he shot a big juvenile bald eagle and decided to stretch the truth a little.

If Washington's eagle is a real bird, though, is there a chance it survived into the present day? Size is hard to estimate without something of known size to compare it to. Is it a gigantic eagle that's really high up or an ordinary eagle at a closer distance? Combine that with Washington's eagle looking so much like a juvenile bald eagle, and there could be a remote population hiding in plain sight.

∼

Audubon's Mistakes

AUDUBON WOULD PROBABLY HAVE STRAIGHT UP MURDERED someone to get his hands on a modern field guide to North American birds. Instead, he had to make his own and some mistakes crept in. Some of them seem to have been "mistakes," where he may have painted a bird he hadn't ever seen.

Selby's flycatcher and Bartram's vireo

Selby's flycatcher was just a female hooded warbler and Bartram's vireo was undoubtedly just a red-eyed vireo. Nothing exciting to see here.

Townsend's finch (or bunting)

This sparrow-like finch with a conical bill was shot by a man named

John Townsend near Philadelphia, Pennsylvania in 1833, and Audubon painted it. The specimen is still in the Museum of Natural History in Washington DC. Ornithologist Kenneth Parks studied it in 1985 and determined it was a dickcissel with unusual plumage. This solution was verified in 2014 when Kyle Blaney, a Canadian birder, photographed an identical bird in Ontario and identified it as an aberrant-plumaged dickcissel.

Cuvier's kinglet

Similarly, this tiny bird has since been identified as a known kinglet. Audubon killed his specimen in Pennsylvania in 1812 and didn't realize it wasn't an ordinary golden-crowned until he picked it up. It looks like a golden crowned kinglet but with an all-red head spot. Since the male golden-crowned kinglet usually has a head spot that's bright yellow with a spot of red in the center, this individual was probably just a golden-crowned with an aberrant head spot.

Carbonated warbler

Audubon supposedly painted this bird from two specimens he shot in Kentucky in 1811. It's a streaky yellow and gray warbler that looks a lot like a blackpoll warbler with yellow areas instead of white. Since blackpoll warblers do show plenty of color variation, including yellowish markings instead of white, some ornithologists now think the birds may have been blackpoll warblers. But the painting's simplicity and some errors of feather arrangements suggest Audubon might not have been painting from a model. Rats chewed up about 200 of his paintings in 1812 and he had to redo them. If he no longer had the original specimen to paint from, he may have repainted this bird from memory and made some mistakes.

Small-headed flycatcher and blue mountain warbler

Audubon had a rival bird artist named Alexander Wilson, although his paintings aren't as good as Audubon's. Wilson painted a blue mountain warbler, so Audubon did too. Audubon painted a small-headed flycatcher, so Wilson did too. But neither bird appears to exist. Audubon claimed Wilson copied his flycatcher from his, Audubon's, painting; and it's probable that Audubon copied Wilson's blue mountain warbler painting so his record would be complete. It's impossible to untangle the confusion in these paintings to learn the truth, but it's likely that both birds were misidentified.

KOAO

T he French painter Paul Gauguin moved to the island of Hiva Oa in 1901, less than two years before he died. Hiva Oa is the second largest island in the Marquesas, a volcanic archipelago in Polynesia in the South Pacific. It's extremely remote, only slightly closer to Mexico than it is to New Zealand, although Hawaii is even closer.

During his short time on Hiva Oa, Gauguin produced quite a few paintings, including a famous one whose title translates to "The Sorcerer of Hiva Oa." It depicts a dancer and magician named Haapuani and represents the fast-vanishing local culture. One of the details of the painting is a bird on the ground next to a small dog. After the previous four chapters you shouldn't be surprised to learn that no one can identify the bird in the painting.

The bird is mostly blue but with green face and wings, and the small dog appears to be biting its wing or back. It's about the size of a chicken, although details aren't especially clear due to Gauguin's painting style. The bill is a brownish-red and is thick and pointed. The eye is the same brown-red.

No one paid much attention to the bird in the painting until a man named Thor Heyerdahl published a book about the Marquesas in 1974. He was most well known for his adventures sailing a balsawood boat, the *Kon-*

Tiki, from South America to the Polynesian islands in 1947. In 1937, though, the
most notable thing Heyerdahl did was
see an unusual bird that he mentioned in
his book.

He wrote that the bird had no wings
and ran extremely fast when he startled
it. It vanished into a thick bank of ferns
and although he tried to catch another
glimpse of it, it was gone. He later said it
was about the size of a long-legged gull.

A French explorer also wrote about
this bird in 1957, although he didn't see it
himself. He said the people who lived on
the island called the bird koao, which
meant "burrow bird" since it was
supposed to hide in burrows. It was about the size of a rooster, purplish in
color with a yellow bill, and while it only had little wings, its legs were long
and it was a fast runner.

By 1979, researchers investigating the koao were told that it had gone
extinct from overhunting, specifically by the French colonizers of Hiva Oa.
Other researchers learned that the bird was supposed to have red eyes and
was the size of a duck.

The ornithologist Jean-Jacques Barloy thought the bird sounded like a
type of rail. Rails are relatively small birds that mostly stay on the ground.
Even rail species that can fly are weak flyers, while many species are flightless. The family is a large one and includes birds like the American coot, the
takahē of New Zealand, the common moorhen that lives throughout much
of western Europe, South Asia, and parts of Africa, and the spotless crake
that's common throughout much of the South Pacific. Barloy suggested in
1979 that the koao might be a spotless crake.

The spotless crake is bluish-gray with reddish-brown back and wings, a
black bill, red eyes, and pale orangey legs that are long for its size. It's shy
and mostly crepuscular, but when it's out in daylight it never goes far from
vegetation where it can hide. It prefers freshwater wetlands but will also
live in forests as long as it has plenty of groundcover for shelter. It eats

insects, worms, crustaceans, and even carrion, as well as plant material like seeds and fruit. It can fly but it would much rather run away from danger.

This description doesn't really fit with what we know of the koao. For one thing, the spotless crake is much smaller than a duck or rooster, smaller even than a crow. It doesn't match the size or coloration of Gauguin's mystery bird or the reports of the koao.

Barloy later decided he was wrong and suggested the koao might be a different type of rail, maybe even an unknown species of takahē. The takahē is dark blue with a greenish back, and its heavy beak and strong legs are red. This is much more similar to the bird Gauguin painted. The takahē was considered extinct until a small population was rediscovered in 1948, and while it's flightless, its ancestors weren't. Like New Zealand, many remote islands—including Hiva Oa—have no native mammals except bats. As a result, many island birds don't need to fly because their predators are other birds like eagles. It's easy to hide from an eagle if you're foraging under cover of thick plants.

Without the bird itself or its remains, though, identifying it is impossible. Gauguin was a post-Impressionist painter who influenced later artists of the avant-garde movement, so his paintings aren't photo-realistic. He was just making art, not illustrating a scientific treatise. The details of his painted bird might not be totally accurate and aren't specific enough to help with an identification. All we know is that the koao looks like a type of rail and doesn't match any known species of bird. So we're back where we started.

But new species of rail keep being discovered in Polynesian islands, most from subfossil remains found during archaeological excavations. In 2007 three new species of extinct rail were described from remains a few hundred years old, while a fourth specimen consisted of only two bones, not enough to identify as a new species. Those two mystery bones were found on Hiva Oa. These findings show that many more species of rail and other birds once lived on the islands, probably driven to extinction by introduced rats and other non-native animals.

A rail described in 1988 from 600-year-old remains, *Porpyrio paepae*, sometimes called the Marquesas swamphen, may be the koao. It lived on Hiva Oa and another nearby island and was closely related to the takahē.

French biochemist Michel Raynal has researched the koao extensively since 1980 and suggests that Gauguin witnessed a dog catch a koao in 1902. That would explain why the dog in his painting is biting the bird's back or wing. If the bird Heyerdahl saw in 1937 was also a koao, we can determine that it was still alive at that time.

The koao may be extinct now, but at least we have a painting of it. That's more than we have for most extinct animals.

PINK-HEADED DUCK

I n the remote wetlands of eastern India and a few nearby countries, an unusual duck was once a shy and elusive resident. The male had pink feathers on his head and neck. The duck built its nest in dense elephant grass and its eggs were almost completely round.

There hasn't been a single confirmed sighting of the pink-headed duck since 1949. Some researchers push this back even farther, to 1935. But people still occasionally report seeing one.

One of the only known photos of the pink-headed duck

The difficulty in knowing whether pink-headed ducks are still alive is that the areas where they once lived are really hard to get to, unless you're a duck. The decline of the species started in the 19th century when British big game hunters would come through and basically just shoot everything that moved. It was already considered rare by the turn of the 20th century, which made hunters even more eager to shoot it so they'd have a rare trophy. Habitat loss and trophy hunting drove it nearly to extinction even if it's not actually already extinct.

Recent expeditions by conservationists and birders haven't found any of the ducks or proof that any are still alive, although a 2017 expedition to Myanmar interviewed locals who said they'd seen the ducks as recently as 2010.

We don't know a whole lot about the pink-headed duck. Researchers think it was a diving duck, but it may have been a dabbler. A dabbling duck tips its body forward, head underwater and tail sticking up, to forage in shallow water, often on plants. A diving duck dives for its food, usually small animals of various kinds. We know the pink-headed duck ate snails and plants, but it probably ate other things too.

A study of a taxidermied pink-headed duck's feathers in 2016 determined that the pink color came from carotenoids, the same pigment that gives the flamingo its pink color. The only other duck with feathers pigmented by carotenoids is the pink-eared duck of Australia, which has a tiny pink spot on each side of its head, but it's only distantly related to the pink-headed duck.

Conservationists, birdwatchers, duck enthusiasts, and people whose favorite color is pink all hold out hope that the pink-headed duck is still alive, hiding its round eggs in clumps of elephant grass far away from humans. Some researchers have even suggested it might be nocturnal, which would explain why it's always been hard to find. It was never much of a duck for moving around, preferring to stay put instead of flying off to other areas.

Hopefully someone will discover a healthy population of pink-headed ducks one day, possibly somewhere no one's even looked yet, and we can protect it and learn about it before it's too late. Once a duck is gone, a duck is gone forever.

~

Imperial Woodpecker

THE IMPERIAL WOODPECKER is the largest woodpecker known. It's over 2 feet long, or 61 centimeters, with a wingspan of probably around 3 feet, or about 92 centimeters—maybe more. It's native to the mountains of western Mexico.

Until the early 1950s, the imperial woodpecker was reasonably widespread. Then lumber companies started logging in the imperial woodpecker's territory. One old man remembered a forester telling locals that the birds destroyed trees, and he even gave them poison to spread on feeding sites. But the imperial woodpecker only feeds and nests in trees that are already dead or dying. It was never a threat to healthy trees. The last confirmed sighting of the imperial woodpecker was in 1956.

No photographs of a living imperial woodpecker exist. Then researcher Martjan Lammertink found mention in a 1962 letter of video taken of a bird in 1956 by dentist and amateur birder William Rhein. Rhein had become reclusive in his old age and moved with no forwarding address at least once, but Lammertink managed to track him down in 1997, when he was in his late 80s. Rhein died in 1999.

Once Lammertink found him, Rhein produced 85 seconds of 16 millimeter movie footage he'd taken back in the 1950s, which showed a female imperial woodpecker hitching up a tree and flying away. From those 85 seconds, researchers learned a lot about the bird, helped by a 2010 expedition that pinpointed the exact location where the footage was shot.

There have been numerous sightings of imperial woodpeckers since the 1950s, but the list is discouraging. The sightings taper off slowly in different areas over the decades. The most recent was 2005, but it hasn't been verified and no photographs were taken.

These days, the areas where imperial woodpeckers once lived are now dangerous to explore due to drug cartels, which grow marijuana and opium poppies in remote clearings with armed guards. If the bird does still exist, at least it's well protected.

IVORY-BILLED WOODPECKER

A lot of people who aren't otherwise into birds have heard of the ivory-billed woodpecker because of the 2004 and 2005 sightings, which were widely reported in the press. This bird may be the most famous of all the entries in this book.

The American ivory-billed woodpecker and the Cuban ivory-billed woodpecker were once considered separate species but are now listed as subspecies. They're big birds, glossy black in color with white markings. The male has a red crest with a black stripe up the front while the female's crest is all black. They need vast areas of undisturbed forest to thrive, something that's in short supply these days.

By the early 20th century, the Cuban ivory-billed woodpecker was already restricted to pine forests in the northeast of Cuba due to habitat loss. By the late 1940s it was rare. In 1956 some small populations were still around, but while conservation was urged, the Cuban revolution in 1959 stopped any conservation progress. The last positive sighting was in 1989. The Cuban government designated the area of its sighting as protected, but no one's seen a bird since.

You probably won't be surprised to hear that the American ivory-billed woodpecker's story is pretty much the same. It's an impressive bird, as much as 21 inches long, or 53 centimeters, with a 2.5-foot wingspan, or 76 centimeters. It likes hardwood swamps and pine forests and was once found throughout the southeastern United States. But as forests were cleared, its habitat grew smaller and more fragmented.

It was thought extinct as early as the 1920s, but then someone spotted a pair in Florida—and promptly shot them as trophies. Another bird was shot in Louisiana in 1932. By 1938, almost the only known ivory-billed woodpeckers were living in a forest in northeastern Louisiana.

To explain what happened then, I need to back up a little. In 1913, the president of the Singer Sewing Machine Company bought almost 83,000 acres of timberland in Louisiana, with further purchases over the next few years that brought the total acreage to about 130,000. That's about 526 square kilometers. He designated the area as a refuge. By this he meant the trees could only be harvested with his permission, mostly for use in his sewing machines, and hunting was not allowed. It was called the Singer Tract, or just Singer by the locals, who continued to use the property as they had for decades—cutting trees for fuel and hunting game for food.

In 1920, Singer got tired of this and offered the property to the Louisiana Fish and Game Department, which hired wardens to enforce trespassing and game laws. The area is frequently called an old-growth forest, but in actuality much of it consisted of abandoned cotton plantations that had been reclaimed by forests.

Interest in the ivory-billed woodpecker had been growing ever since it was rediscovered in the 1920s after its supposed extinction. In 1935, Cornell University sent a team of researchers to the Singer Tract to look for the birds. The team brought film and recording equipment instead of guns. They found the woodpeckers and took pictures and sound recordings.

The expedition was so successful that one of its members returned in 1937 to study the ivory-billed woodpecker for three years. Also in 1937, Singer sold 6,000 acres to a lumber company, and in 1939 he sold timber rights to the rest of the acreage to the Chicago Mill and Lumber Company.

In 1940, the Audubon Society convinced a Louisiana senator to introduce a bill to establish a national park protecting what remained of the Singer Tract. There was no money to fund the bill, so John Baker, an Audubon Society member, got pledges of support from the heads of the U.S. Forestry Service, U.S. Fish and Wildlife Service, and the National Park Service. He even got an endorsement from President Roosevelt for the bill.

The governor of Louisiana pledged $200,000 for the purchase of the land, and in 1942 the head of the War Production Board confirmed that clearcutting the Singer Tract was not essential to the war effort. Governors of the neighboring states of Tennessee, Arkansas, and Mississippi sent a joint letter to the Chicago Mill and Lumber Company asking that they release their lease on the remaining timber.

Senator Ellender reintroduced the bill in 1942 with private funding taken care of, but it failed to get out of committee. And in December of 1943, the Chicago Mill and Lumber Company basically said they had no interest in conservation. They clearcut the remaining land. The last ivory-billed woodpecker was dead by 1944.

I wish I could tell you that the Chicago Mill and Lumber Company foundered and that its president choked to death on a bite of roast chicken. Unfortunately, the company did very well selling timber in the post-war boom. In 1965 the remaining Singer acreage was bought by a company in Chicago, and the lumber company leased the woodlands to private hunting clubs for a few years. Then they bulldozed and burned what was left of the timber to make way for soybean crops. And no, the locals were really not happy about all this.

In 1980, what was left of the area was finally bought by the state. The Tensas River National Wildlife Refuge was dedicated in 1998 and looks like a nice place now, but its only ivory-billed woodpeckers are a pair of stuffed specimens on display.

Of course there were numerous sightings of the bird in different areas, but they didn't amount to much. In 1971 someone took two grainy photos that might have been of an ivory-billed woodpecker. In 1999 a forestry student sighted a bird but didn't get a picture. Things like that. Then, in 2004, sightings started trickling in from Arkansas.

It started quietly enough. A kayaker posted online about seeing an unusually large woodpecker in a wildlife refuge. A team led by the Cornell Laboratory of Ornithology conducted a secret intensive search of the area—secret so the place wouldn't be inundated by birdwatchers.

That search resulted in more than a dozen sightings, possibly all of the same bird. The team even managed to catch a bird on video on April 25, 2004. Quietly, secretly, the Nature Conservancy and Cornell University bought up some of the land in the area to add to the wildlife refuge, just in case.

The sightings were made public in early 2005, when an article appeared in the journal *Science*. Cornell declared the bird rediscovered instead of extinct.

Unfortunately, the four-second video taken in 2004 is blurry. William Rhein's 1956 footage of the imperial woodpecker [see the previous chapter] is a lot clearer and he shot it from the back of a mule. It's impossible to determine from the 2004 footage whether the bird is an ivory-billed woodpecker or not. Skeptics believe it might be a pileated woodpecker, a crow-sized bird with similar markings that isn't actually very closely related to the ivory-billed woodpecker.

The exchange of papers got heated, to say the least. Birders split into two camps: those who believed the sightings were of ivory-billed woodpeckers and those who believed the sightings were of pileated woodpeckers.

The problem is, while the video evidence isn't very persuasive, the audio is. The ivory-billed woodpecker's calls were well documented by the 1935 expedition, and the 2004 and 2005 recordings seem to be of the same type of bird.

The 1935 recording was taken very close to the birds. In order to compare it with the new recording, the team took the original recording to the same area and played it back in the distance. The audio sounds very similar to the modern recordings.

Further searches for ivory-billed woodpeckers turned up nothing. By 2010 the excitement had died down and searches were called off, although it's been a boon to Arkansas's tourist industry. Birders and conservationists continue the search, though, and occasionally record what might be the bird's call.

It's always possible the ivory-billed woodpecker still hangs on in various areas. The problem is whether any remaining populations have enough genetic diversity to survive even in ideal conditions in this point.

Here's a reminder that the pileated woodpecker is doing just fine. It's not as big as the ivory-billed woodpecker but it's a large, handsome bird common in forested areas of eastern North America and parts of the west coast. Maybe you won't ever get to see an ivory-billed woodpecker, but you can definitely appreciate the pileated woodpecker.

CHICKCHARNEY

The Bahamas is a country made up of over 700 islands, many of them tiny, located roughly between the Florida peninsula and the much larger island of Cuba. These days it's famous for sunny beaches and warm waters. Tourism is a big part of its economy and lots of people take cruises to the Bahamas. But between about 500 years ago and 200 years ago, the Bahamas was a terrible place. The native people of the area, called the Lucayan, were enslaved by the Spanish and forced to work on plantations under horrific conditions. Most of them died. The British took over the islands around the mid-17th century, bringing enslaved people from Africa to work the plantations.

Also during this time, pirates treated the area as a haven, leading eventually to one really good *Pirates of the Caribbean* movie and a lot of lackluster sequels, although this is perhaps a little off topic. In 1807 the British came to their senses and abolished the slave trade, although they didn't actually abolish slavery until 1834. British ships sometimes attacked slave ships and rescued the captives on board. Many of the captive people were brought to the Bahamas, where they made new homes. Freed and escaped slaves made their way to the Bahamas too, where they could live in relative peace.

The largest of the islands that make up the Bahamas is called Andros Island, although it's technically a collection of three main islands and some smaller ones that are all quite close together, protected by a barrier reef. It's the only island in the Bahamas with a freshwater river and many animals found on Andros Island live nowhere else. There used to be even more native animals, before the forests of Andros were chopped down.

Andros Island was once supposed to be home to a creature called the chickcharney. It was sort of a bird, sort of a goblin. It was about 3 feet tall, or 92 centimeters, with big round eyes—possibly only one eye in the middle of its face. It was covered with hairy feathers and could turn its head almost all the way around. Some versions of the story say it had a long prehensile tail that it used to climb trees. It was supposed to live in the pine forests and make its nest in trees that were so close together that the branches touched near the top.

The chickcharney would sometimes play tricks on people, but if people treated it with respect and left it alone, they would have good luck. If they bothered it, not only would they have bad luck, sometimes the chickcharney would grab the person and twist their head around backwards. The best way to keep the chickcharney from bothering you was to carry brightly colored cloth or flowers when you went into the woods.

At first glance, the story of the chickcharney doesn't seem very believable. But as it happens, Andros Island used to be home to a flightless owl that sounds a lot like the chickcharney (minus the head-twisting part).

The Andros Island barn owl stood over 3 feet tall, or about 92 centimeters, with long legs. It was a burrowing owl that nested in holes beneath pine trees. It probably went extinct in the 16th century when the pine forests on Andros Island were felled, but occasionally people still report seeing the chickcharney. So while it's a slim chance, maybe a small population of the owl is still hanging on.

GIANT PENGUIN

T he emperor penguin lives in Antarctica and can grow over 4 feet tall, or 130 centimeters, which is just ridiculously large. It can also weigh up to 100 pounds, or 45 kilograms. In other words, it's as big as a small person. It's a strong swimmer and its deepest recorded dive was well over 1,800 feet, or 565 meters. That's *whale* diving depth.

But the emperor penguin isn't the biggest penguin that ever lived. Anthropornis went extinct around 33 million years ago and it was a penguin that was actually the height of a tall human, some 6 feet tall, or 1.8 meters. It lived off the coast of what is now New Zealand and Antarctica. The New Zealand giant penguin lived around the same time as Anthropornis and was a bit over 5 feet tall, or 1.6 meters, but probably weighed more. Neither were direct ancestors of modern penguins, but they probably looked and acted very similar.

A newly discovered giant penguin, also from New Zealand, lived much earlier than the others. It was almost 5 feet tall, or 1.5 meters, and well adapted to the water 61 million years ago. Some researchers hypothesize that penguins had already begun evolving when dinosaurs were still alive, and that they survived the extinction event 66 million years ago.

Back in the 1920s and 30s, when fossils of giant penguins were first described, they caught the public's imagination. Giant penguins appeared

in science fiction of the day, including Jules Verne and H.P. Lovecraft. Then, starting in February of 1948, people in Florida began finding enormous three-toed tracks in sand on a few beaches and along the Suwannee River. The footprints were over a foot long, or 35 centimeters, and the animal's stride was measured at between 4 and 6 feet long, or 1.2 to 1.8 meters.

Cryptozoologist Ivan Sanderson examined the tracks in November of 1948. After weeks of study he reported gravely that they'd been made by a penguin 15 feet tall, or 4.5 meters.

It turns out, though, that it was all a hoax. Two men, Tony Signorini and Al Williams, had made gigantic iron feet they could wear as great big shoes. They walked in the sand overnight leaving trails of monstrous tracks ready to be discovered by beachcombers. Each foot weighed about 30 pounds, or 13.5 kilograms, and Signorini used the weight to swing along in a sort of controlled bound that made his stride remarkably long without too much effort. They actually intended the tracks to be taken for dinosaur or sea monster footprints, but a giant penguin was even better.

Williams died in 1969 but Signorini didn't come clean about the hoax until 1988. He still has the feet.

PART THREE
FRESHWATER MONSTERS

You won't learn about Nessie in this section, but you'll learn about some other lake and river monsters like the freshwater seahorse of Asia and South America, the inkanyamba of Africa, the lagarfljót of Iceland, and more!

BURU

In northeastern India, in a remote swampy area in the foothills of the Himalayas, a group of people called the Apa Tani lived in a broad valley. They were expert rice farmers and although they had no written language, they kept a detailed oral history.

In 1945 and 1946, two British men, Charles Stonor and J.P. Mills, visited the Apa Tani. Mills was an anthropologist who wanted to learn more about how the people lived and what their traditions were. Stonor was many things, including a zoologist, an anthropologist, and an explorer, although it's not clear how formal his education was. He published several scholarly articles in the field of anthropology and one book about the Yeti that was less scholarly but way more popular, with fourteen editions published between 1955 and 1958.

But we're not talking about the Yeti. Mills and Stonor learned about a much different mystery animal from the Apa Tani, called the buru.

According to the Apa Tani, their ancestors migrated to the Ziro valley along two rivers. Accounts of their migration match up with actual places with a high degree of accuracy even though the migration took place many centuries ago.

When the Apa Tani reached the valley, it was largely flooded with a swamp and lake. The buru lived in deep water in the lake but occasionally came to the surface, stuck its head above water, and made a noise translated by Mills and Stonor as a hoarse bellow. Occasionally a buru would nose through the mud in shallower water, and it was described as eating mud, not fish. It was about 4 meters long, or a bit over 13 feet, and was dark blue blotched with white, with a white belly.

The Apa Tani drained much of the swamp and lake to create more rice farmland, and on four occasions a buru was trapped in a pool of deeper water. The Apa Tani killed the burus trapped this way and buried their bodies, and the location of the buried burus are still known, or were in the 1940s. The Apa Tani reported that there were no more burus in the valley.

The story of the buru interested Charles Stonor. When he was traveling near the Apa Tani's valley a while later, he met a few members of another tribe and asked them if they'd ever heard of the buru. To his surprise, they said they not only knew about the buru, it lived in a swampy valley not too far away, called Rilo, home to a group of people called the Dafla. Naturally Stonor visited the valley as soon as he could.

The Dafla confirmed that the buru did indeed live in the swamp, which was about 50 miles away from the Apa Tani's valley, or 80 kilometers, and separated from it by tall ridges.

The buru of the Rilo swamp was a little different from the buru of the Apa Tani. It was described as the size of a man and black and white in color, with a head as large as a bison's but with a longer snout, and with a pair of small backward-pointing horns. It lived underwater and only came to the surface briefly, accompanied by considerable splashing, but was only seen in summer when the swamp flooded and became a lake. No one in the Rilo valley had ever seen a buru up close.

In early 1948, rumors of a living dinosaur in the remote areas of northern Assam reached a British man named Ralph Izzard. Izzard was a foreign correspondent for the London *Daily Mail* and he smelled a story. He approached Stonor to ask if he wanted to undertake a small expedition to look for the so-called dinosaur, the buru. Stonor agreed, and in April 1948 the expedition headed to Rilo with a photographer.

They watched the swamp daily for months and saw exactly zero burus. Since it had never been an animal the villagers paid much attention to, no one had realized it was gone. This sounds absurd until you realize that the village had only been settled about a decade before. Many trees had been felled in the process, which increased erosion so that the swamp had silted up considerably and was no longer very deep even at full flood. It's probable that the burus died due to these changing conditions, especially if they hadn't been very numerous to start with.

The expedition returned to civilization only to find that rumors of the buru hunt had leaked. The papers were full of reports of a 90-foot, or 27-meter "dinotherium" sighted in the jungle. That was kind of embarrassing since they hadn't seen anything mysterious at all.

So, what was the buru?

Izzard and Stonor were convinced it was a reptile of some kind. Let's look more closely at the Apa Tani's description of the animal to get a clearer idea of what it might be.

The buru had a snake-like head with a long snout that was flattened at the tip. Its eyes were deep-set and it had three hard plates on its head that helped it burrow in the mud. Its teeth were small and flat except for a larger pair in the upper and lower jaws. Its body was as big around as a man could reach and it had a long tail that wasn't very pointed but that had a fringed lobe all around it. Its skin was fishlike but without scales, and it had three rows of short spines along the back and sides.

It also had limbs, but while most people of the village called them short legs with claws used for digging, one old man stubbornly refused to describe them as legs. Mills, the anthropologist, found this confusing because he assumed the old man, a priest named Tamar, was talking about a reptile.

Tamar said the buru had a body like a snake, without legs but with two pairs of appendages as thick as a man's arm along its sides. He insisted that these were definitely not legs.

I wonder if he was trying to explain something he probably didn't fully understand, lobed fins.

There's one animal, a strange type of fish, that fits almost all of the details of the buru. Lungfish live in Africa, South America, and Australia, with the Australian species retaining more primitive characteristics than are found in the others. Instead of ordinary fins, the Australian lungfish has lobed fins that resemble stubby legs with a frill around them.

Izzard rejected the idea that the buru was a lungfish, because, he wrote, "no known fish would expose itself above water, for no practical purpose, for such a length of time."[1] But that's exactly what the lungfish does. The lungfish breathes through its gills most of the time like any other fish but it also has lungs. When it surfaces for a breath, it makes a big splash and a big grunting gasp.

Stories of the buru eating mud also sound like the lungfish, which eats crustaceans and snails it digs out of the mud. Even more interesting, the Dafla reported that their black-and-white buru was only seen in summer when the swamp flooded to form a lake. This matches what we know of other lungfish, which generally need extra oxygen during spawning season when they're most active and will gulp air to replenish their lungs. During the rest of the year, the buru would probably have spent most of its time hidden in the swampy water.

There are a few details that don't match with a lungfish, of course. Lungfish don't have forked tongues, for one thing. Many cryptozoologists think this forked tongue points to a type of monitor lizard, but while some monitor lizard species do spend a lot of time in the water, notably the widespread Asian monitor lizard, the buru is described as being exclusively aquatic. Monitor lizards are also very lizardy, with large, strong legs that it can walk around on, and it eats fish, frogs, rodents, and other animals.

It's frustrating to think that if Stonor had heard about the Rilo swamp buru just a few years earlier, he might have encountered a live one. Still, it hasn't been all that long since the last sightings took place. There might be remains to be found buried in the swamp.

There might even be living burus hidden in the remote swamps and lakes of northeastern India.

AUSTRALIAN LUNGFISH

In 1869, a farmer visiting the Sydney Museum asked why there were no specimens displayed of a big olive-green fish from some nearby rivers. The curator, Gerard Krefft, had no idea what the guy was talking about. No problem, the guy said, or probably no worries, he'd just get his cousin to send the museum a few. Not long after, a barrel full of salted greenish fish that looked like big fat eels arrived and Krefft set about examining them.

When he saw the teeth, he practically fainted. He'd seen those teeth before—in fossils several hundred million years old. No one even knew what fish those teeth came from, but here they were in fish that had been pulled from a local river only days before.

The Australian lungfish doesn't have ordinary teeth. Instead, it has four tooth plates or combs that resemble regular teeth that have fused together. Its skull is also very different from all other fish, possibly because of its feeding style. It crushes prey with its tooth combs, so its skull has to be able to withstand the force of its own bite. Other lungfish species share this trait to some degree, but with modifications that appear more recent.

The Australian lungfish lives in slow-moving rivers and deep ponds and hunts using electroreception. Larger ones mostly eat snails and crustaceans, while smaller ones also eat insect larvae and occasionally small fish. It can grow up to about 5 feet long, or 1.5 meters. Its body is covered with large overlapping scales and its four fins look more like flippers or paddles. Its tail comes to a single rounded point. In short, it looks superficially like a coelacanth, which is not a big surprise because it's related to the coelacanth. While the Australian lungfish doesn't actually get out of the water and walk on its fins, it does stand and sometimes walk on them underwater.

Unlike other lungfishes, the Australian lungfish has only a single lung instead of a pair. Most of the time it breathes through its gills, but at night when it's active, or during spawning season or other times when it needs more oxygen, it surfaces periodically to breathe. When it does so, it makes a distinctive gasping sound. During droughts when its pond or river grows shallow, an Australian lungfish can survive when other fish can't. As long as its gills remain moist, it can survive by breathing air through its lung. But unlike other lungfish, it doesn't aestivate in mud.

The Australian lungfish hasn't changed appreciably for the last 100 million years. The only real change it exhibits from its ancestors 300 million years ago is that it's not as big, since they grew some 13 feet long, or 4 meters. Lungfish used to be widespread fish that lived in freshwater back when the world's continents were smushed together in one supercontinent called Pangaea, some 335 million years ago. When Pangaea began to break up into smaller continents about 175 million years ago, various species of lungfish remained in different parts of the world. Now we've only got six species left...maybe.

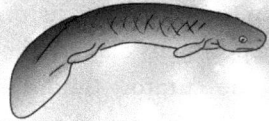

FURRY FISH

Sometimes you'll see a mounted fish that has fur, usually decorating a restaurant. Fur-bearing trout are jokes by taxidermists, who usually attach rabbit fur to a stuffed fish.

But some cultures have stories about fish with hair. This includes the Japanese story of big river fish with hair on their heads like people, although since these fish are supposed to come out of the water at night to fight and play, they're probably not actual fish. There's also an Icelandic legend about an inedible trout with fur that shows up in rivers where people are not being nice enough.

Could these stories be based on a real animal? Are there any fish that grow fur or hair?

Mammals are the only living animals that grow actual hair from specialized cells, but lots of animals have hair-like coverings. Baby birds have downy fuzzy feathers that look like hair and many insects have hairlike structures called setae, made of chitin, that make them look furry.

Some fish grow hairlike filaments that help camouflage them among water plants and coral. The hairy frogfish lives in warm, shallow waters, especially around coral reefs, and grows to about 8 inches long, or 20 centimeters. The hairlike filaments that cover its body help it blend in among seaweed and anemones. It's usually brownish-orange or yellowish, but it can actually change its color and pattern to help it blend in with its surroundings. This color change doesn't happen fast, though. It takes a few weeks.

Like other frogfish, the hairy frogfish it has a modified dorsal spine called an illicium with what's called an esca at the end. In deep-sea species of anglerfish, the esca contains bioluminescent bacteria, but in the hairy frogfish it just looks like a worm. The fish sits immobile except for the illicium, which it twitches around. When a fish or other animal comes to catch what looks like a worm swimming around in the water, the frogfish gulps the animal down. Like other frogfish species, the hairy frogfish has large, strong pectoral and pelvic fins that it uses to walk across the sea floor instead of swimming.

Another fish that looks like it has hair is called the hairyfish. The hairyfish barely grows more than 2 inches long, or 5.5 centimeters. It eats copepods and other tiny crustaceans that live near the ocean's surface and it's covered with small hairlike filaments. Its close relations are equally small fish called tapetails because the tail fin has a narrow extension at least as long as the rest of its body, called a streamer.

The tapetail was described in 1956 but scientists were confused because no one had ever found an adult tapetail, just young ones. It wasn't until 2003 that a team of Japanese scientists discovered that the DNA of tapetails matched the DNA of a deep-sea fish called the flabby whalefish. There are lots of whalefish species, but the largest only grows to about 16 inches long, or 40 centimeters. It looks very different from its larval form, with loose skin without scales or hair-like filaments or the tail streamer.

Even after researchers figured out that the tapetail and hairyfish are larvae of whalefish, there was still another mystery. All the whalefish ever found were females. Where were the males? Finally they identified yet another deep-sea fish called a bignose fish as the male of the species. The bignose fish has a huge liver but its mouth doesn't go anywhere—it doesn't have a throat or stomach. It gets its name from a bulge on its snout that gives it a keen sense of smell.

It turns out that after a larval whalefish develops into an adult, the male doesn't need to eat. It lives off the fat and nutrients stored in its huge liver and uses its sense of smell to find a female in the depths of the ocean. The female remains a carnivore, eating any small animals it can catch, and it often migrates at night from the deep sea to nearer the surface, then returns to the depths during the day.

The hairy frogfish and the hairyfish are both rarely seen marine fish, so it's doubtful that they inspired the legends of furry fish. Are there hairy-looking freshwater fish?

There is a disease called cotton mold that infects fish and makes them look like they have white or grayish spots of fur. Saprolegnia is the name of the mold, which lives in water and can infect fish in the wild and in aquariums. It mostly prefers cold freshwater and usually infects fish that are already injured. It spreads across the fish's skin and makes it look fuzzy, and eventually it kills the fish. Salmon and trout are common targets of this mold, which may be the source of the Icelandic story.

As for the Japanese story about the hairy fish creatures that come out of the river at night, zoologist Karl Shuker suggests the legend may be based on sightings of the northern fur seal. While seals are mammals, not fish, they do look superficially like fish, and while seals also usually live in the ocean, they occasionally stray into rivers.

Next time you go on a fishing trip or just hang out in a boat, keep an eye out for fish with fur. If you see one, pet it and report back to me about how soft its fur is.

WHITE RIVER MONSTER

The White River originates in the mountains of northwestern Arkansas and flows from there through Missouri, then back into Arkansas where it joins the Mississippi River. In 1915 a man near the small town of Newport, in the central Ozarks region of Arkansas, saw an enormous animal with gray skin in the river.

A few other people saw it too but it wasn't until July of 1937 that the monster returned, and this time a lot of people saw it. News of it hit the local papers and spread throughout the country, and people started showing up to look for it.

Estimates of the monster's size varied quite a bit. A man named Bramlett Bateman, who owned a lot of the farmland along that stretch of the river, was quoted in several newspaper articles. He described the monster as being the length of three cars in one article, but in another his estimate was smaller, only 12 feet long, or 3.7 meters, and 4 or 5 feet wide, or 1.2 to 1.5 meters. It doesn't seem that he or anyone else got a really good look at it.

It was described by numerous people as being gray-skinned. Bateman said it had a catfish-like face too. The only description given in a *New York Times* article from July 23, 1937 is this:

Half a dozen eye-witnesses...reported seeing a great creature rise to the surface at rare intervals, float silently for a few minutes and then submerge, making its presence known only by occasional snorts that bubbled up from the bottom.

Another article quotes Bateman as saying he saw the monster "lolling on the surface of the water."

Bateman decided he was going to blow the monster up with dynamite. The local authorities said, uh no, you cannot just throw dynamite into the river. Other people brought machine guns and other weapons and patrolled the river looking for the monster. A plan to make a giant net and catch the monster petered out when people discovered that making and deploying a net that big is expensive and difficult.

The monster was mostly reported in an eddy of the river that stretched for about a mile, or 1.6 kilometers, and was unusually deep, about 60 feet deep, or 18 meters. The river is about 75 feet wide at that point, or 23 meters. The Newport Chamber of Commerce hired a diver from Memphis named Charles B. Brown, who brought an 8-foot, or 2.4 meter, harpoon with him when he descended into the river. He didn't find anything but the tourists watching him had fun.

Suggestions as to what the monster might be ranged from a sunken boat that sometimes bobbed briefly to the surface to a monstrous catfish. Many people were convinced it was a huge fish of some kind, especially an alligator gar.

Eventually sightings tapered off and the excitement died down until June of 1971, when it started being seen again. Again the size estimates were all over the place, with one witness saying it was the size of a boxcar, which would be about 50 feet long, or 15 meters, and 9 feet wide, or 2.8 meters. Another witness said it was only 20 feet long, or 6 meters. Some witnesses said it had smooth skin that looked like it was peeling all over, had a bone sticking out of its forehead, and it made sounds that one witness described as similar to both a horse's neigh and a cow's moo. On July 5, 1971, three-toed tracks 14 inches long, or 36 centimeters, were also found on an island together with crushed plants that showed a huge animal had come out of the water.

This time, at least, no one tried to dynamite or even net the monster. Instead, in 1973 Arkansas passed a law creating the White River Monster

Refuge along that section of the river, to protect the monster. But no one has seen it since.

There is a photo of the monster taken in 1971, but it's a blurry Polaroid that was reproduced in a newspaper and the original lost. The photo was taken by a man named Cloyce Warren, who was out fishing with two friends. Warren said it had a row of spines along its back.

Obviously people are seeing something in that part of the White River, but it's reportedly so big that if there was a population living anywhere in the river, it would be spotted all the time. It might be an animal that only sometimes strays into the White River and actually lives in the much larger Mississippi River—or even in the Gulf of Mexico, where it sometimes swims upriver.

People have made plenty of suggestions over the years. One suggestion is that it's an elephant seal. The northern elephant seal is an enormous animal, although it's nowhere near 50 feet long. The male is much larger than the female, up to 16 feet long, or 4.8 meters, and bulky with blubber that keeps it warm when diving deeply for food in the Pacific Ocean where it lives. But the endangered elephant seal *only* lives in the Pacific, which is separated from the Gulf of Mexico by a whole lot of the North American continent.

Another suggestion is a manatee—specifically the Florida manatee. In the winter it mostly lives around Florida but in summer many individuals travel widely. It's sometimes found as far north as Massachusetts along the Atlantic coast and as far west as Texas in the Gulf of Mexico.

The manatee is large, up to 15 feet long, or 4.6 meters, with females being somewhat larger than males. Its skin is gray but since it moves slowly, it can look mottled in color due to algae growing on its skin. It sometimes also has barnacles stuck to it the way some whales do. It has a pair of front flippers with three or four toenails, no hind legs, and a paddle-like tail. It eats plants and only plants, and is completely harmless to humans, fish, and other animals. Because it moves slowly and spends a lot of time at the surface, since it's a mammal and has to breathe air, it's vulnerable to being injured by boats.

The Florida manatee nearly went extinct in the 1970s, when only a few hundred individuals remained. It was listed as an endangered species and after a lot of effort by a lot of different conservation groups, it's now only considered threatened. So while people might recognize a manatee these days, back in the 1970s it was practically unknown everywhere except southern Florida since it was so rare. And in the decades before 1971, people didn't travel as much and didn't know much about increasingly rare animals that didn't live in their particular part of the world.

In other words, it's completely possible that people from Arkansas would see a manatee in 1915, 1937, and 1971 and not know what it was. But could a manatee really travel that far from the ocean and survive?

The Mississippi River empties into the Gulf of Mexico in Louisiana in the United States. Texas is to the west of Louisiana and to the east are Mississippi, Alabama, and Florida. In other words, it's well within the known range of the Florida manatee. Manatees are known to sometimes travel up the Mississippi. This happened most recently in October of 2016 when a manatee traveled as far as Memphis, Tennessee before it was found dead in a small lake connected to the river. That's a distance of 720 miles, or 1,158 kilometers, and that was with wildlife officials trying to capture it to return it to the Gulf. That same year a manatee also traveled as far as Rhode Island along the Atlantic coast. Memphis is actually much farther up the Mississippi than the White River is, so if the manatee had branched off into the White River it might have led to new sightings of the White River Monster.

The manatee can live in fresh water perfectly well. One species, the Amazonian manatee, is a fully freshwater animal that never leaves the South American rivers where it lives. But despite its size, the manatee doesn't have a lot of blubber or fat to keep it warm. The farther away it travels from warm water, the more likely it is to die of cold.

While an errant manatee might explain some White River Monster sightings, it doesn't fit all of them. Other animals from the Gulf of Mexico sometimes find their way up the Mississippi too. It's a huge river, and since an ocean animal doesn't understand what a river is, it doesn't know it's never going to reach the ocean again unless it turns around. Most marine animals can't survive for long in fresh water, but some can tolerate fresh water. That's the case for the bull shark.

In 1937, the same year the White River Monster was spotted for the second time, a 5-foot bull shark, or 1.5 meters, was caught in Illinois, which is even farther upstream from the Gulf of Mexico than Tennessee and Arkansas. Bull sharks live throughout much of the world's oceans in warmer water near coasts and are often found in rivers and lakes, although they don't live as long in freshwater as they do in salt water. The largest bull shark ever measured was 13 feet long, or 4 meters, so a large one is about the size of a manatee.

Occasionally a dolphin travels up the Mississippi River, but marine dolphins can't survive for long in freshwater and will die soon if they can't make their way back to the ocean. A dolphin in freshwater starts to develop skin lesions and then the skin begins to peel, leading to bacterial infection and death. Remember that some witnesses in 1971 described the White River Monster as a gray animal with peeling skin.

Nine different species of dolphin and many species of whale live in the Gulf of Mexico. Of those, only the bottlenose dolphin lives close to the coast and is usually the species that accidentally travels into fresh water and can't find its way out. The bottlenose dolphin is a little smaller than the manatee, up to about 13 feet long, or 4 meters.

1971 was an active hurricane year, including the category five Hurricane Edith that killed 37 people in mid-September. Marine animals that can travel quickly, like dolphins and sharks, flee to calmer waters when a hurricane approaches. That usually means out to sea, but it wouldn't be out of the question for a frightened dolphin or other large marine animal to swim into the Mississippi River by accident ahead of a hurricane, especially a hurricane as big as Edith.

Another possible identity for the White River Monster is one that was suggested in 1937, the alligator gar. It's a freshwater fish that lives throughout the Mississippi River and other rivers and lakes in the southern United States and parts of northern Mexico. The alligator gar gets its name because of its toothy jaws, which do resemble an alligator's, and it can grow up to 10 feet long, or 3 meters. It has gills like other fish but it can also breathe air through its swim bladder, which is lined with lots of blood vessels that absorb oxygen. Every so often an alligator gar will come to the surface and gulp air to replenish the oxygen in its swim bladder, so it would be seen at the surface briefly but periodically as was described by many witnesses. This is also the case for the manatee and dolphin, which breathe air.

The alligator gar is an ambush predator, which means it waits in the water without moving much at all until an animal approaches. Then it shoots forward and grabs it. It mostly eats small fish, invertebrates of various kinds, and waterfowl like ducks.

Alligator gar

Another possibility for the White River Monster's identity is the gulf sturgeon. It's a subspecies of the Atlantic sturgeon that lives in the Gulf of Mexico, although it's also known from various rivers in the southeastern United States. The reason it's found in rivers is that the gulf sturgeon is anadromous, the term for a fish that migrates from the ocean into freshwater to spawn. The salmon is the most famous anadromous fish, fighting its way upriver to spawn and then die. In the case of the gulf sturgeon, it hatches in freshwater and lives there for the first two years or so of its life before making its way downstream to the ocean. Then it returns to freshwater to spawn every spring, usually the same river where it was hatched, and goes back to the ocean in autumn.

The gulf sturgeon fits a lot of the descriptions of the White River Monster sightings. It's covered with five rows of scutes that project from the back and sides in a sort of low sawtooth pattern, which fits the row of spines that Cloyce Warren reported seeing in 1971. Its elongated snout has sensory barbels like a catfish, which matches Bramlett Bateman's 1937 description of the monster having the face of a catfish. It's gray, gray-green, or brownish in color with a lighter belly, and it can grow up to 15 feet long, or 4.5 meters, although most are about half that length.

In the winter the gulf sturgeon lives just off the coast in shallow water, where it's a bottom feeder. It sucks up invertebrates from the sea floor, feeling for them with its barbels. When it's fat and healthy in summer, it migrates upriver in groups, but occasionally one gets separated from its group and finds its way into a stretch of water by itself. Sturgeons do sometimes jump out of the water, especially in summer—as much as 6 feet out of the water, or 1.8 meters. No one's sure why. Also during the summer, the sturgeon makes a sound like a creaky hinge.

Finally, the White River Monster might have been an American alligator. A big male gator can grow over 15 feet long, or 4.6 meters, with reports of much larger individuals. The gator is usually dark in color but can be gray, and it has rows of bony scutes down its back. Unlike the other animals we've talked about, including the manatee, the alligator also comes out of the water, so could be responsible for the three-toed footprints found in 1971. Then again, even the biggest gator doesn't have a 15-inch footprint, and it has five toes on its front feet and four on its hind feet.

I think it's probable that the White River Monster sightings are of more than one type of animal. While we can make an educated guess as to which animals might have been spotted and misidentified, we can't know for sure. There's a possibility that something else occasionally swims up the Mississippi from the Gulf and into the White River, something unexpected and maybe even unknown to science.

Hopefully, next time the White River Monster appears, someone gets a really good look at it and some good pictures.

TENNESSEE RIVER MONSTER

The earliest report of the Tennessee River Monster is from April 1878 in the Chattanooga *Daily Times*, an account from an old resident about river monster sightings from earlier that century. The first sighting by a white settler is reportedly from 1822, when a man named Buck Sutton was fishing and sighted the monster. The next sighting was near the same area five years later, when a man named Billy Burns saw the monster while crossing the ferry. Jim Windom was fishing in 1829 when he saw it. All three men died the summer after their encounters, although subsequent sightings (including 1836 and 1839) didn't lead to anyone's death.

The sightings all apparently took place in a part of the Tennessee River near Chattanooga, now dammed to form Chickamauga Lake. At the time the river there was relatively sluggish and shallow, with many shoals.

The monster was described as serpent-like and about the length of a canoe, or around 20 to 25 feet long, or 6 to 7.6 meters. At least one report says it had a doglike head. Billy Burns reported that its belly was yellow and its back was blue. The most interesting detail comes from at least two reports, that of a tall black fin on its back that stood at least 18 inches high, or 45 centimeters, or possibly 2 feet high, or 61 centimeters, above the water.

The Tennessee River has its share of unusual animals, from tiny freshwater jellyfish to the paddlefish, a filter feeder with an elongated rostrum. In shallow water the tail fins of fish can show above the surface higher than the dorsal fin, but not 2 feet out of the water. No freshwater fish with such a large and prominent dorsal fin lives in North America.

It's possible that on rare occasions, a bull shark could find its way into the Tennessee River. The Tennessee is a tributary of the Ohio River, which in turns flows into the Mississippi, which then empties into the Gulf of Mexico. While bull sharks do occasionally swim up the Mississippi, no genuine sighting of one in the Ohio or Tennessee rivers has ever been documented. It's not impossible, though. An exceptionally large bull shark can grow up to 13 feet long, or 4 meters, and it prefers shallow water. Tennesseans in the early 19th century would have no knowledge of sharks and might consider it a monster, not an ordinary fish.

However, the sightings specify a snake-like creature, not a fish. The Tennessee River might once have been home to a large, slender fish with a tall dorsal fin, one that was already rare in the early 19th century and which went extinct soon after.

It's also possible that the story was just a newspaper hoax, written to fill space on a slow news day. The article from 1878 was a "contribution...from an old citizen of Chattanooga" who was not named, talking about events that took place more than fifty years before. In 1885 another newspaper, the Chattanooga *Daily Commercial,* ran a nearly identical article—obviously taken from the 1878 one, often word-for-word—that claims the reporter heard the story "yesterday while listening attentively to the conversation of one of Chattanooga's oldest citizens."

We may never know what the strange Tennessee River animal was—but just to be on the safe side, I'm staying in the boat.

FRESHWATER SEAHORSES

Seahorses are definitely marine animals. That means they only live in the ocean. But rumors of freshwater seahorses have been around for decades.

Freshwater seahorses are supposedly known in the Mekong River in Asia and Lake Titicaca in South America. Sometimes you'll see reference to the scientific name *Hippocampus titicacanesis*, but that's not an official name. There's no type specimen and no published description. Another scientific name supposedly used for the Mekong freshwater seahorse is *Hippocampus aimei*, but that's a rejected name for a seahorse named *Hippocampus spinosissimus*, the hedgehog seahorse. It does live in parts of the Indo-Pacific Ocean, including around Australia, especially in coral reefs, and sometimes in the brackish water at the Mekong River's mouth, but not in freshwater.

On the other hand, there's no reason why a seahorse couldn't adapt to freshwater living. A few of its close relatives have. Most pipefish are marine animals but some have adapted to freshwater habitats. This includes the black-striped pipefish, which is found off the coasts of much of Europe but which also lives in the mouths of rivers. At some point it got introduced into the Volga River and liked it so much it has started to expand into other freshwater lakes and rivers in Europe. In the world of aquarium enthusiasts, freshwater pipefish are actually sometimes called freshwater seahorses.

The pipefish looks like a seahorse that's been straightened out. While it does have bony plates like a seahorse, it's a more flexible fish than the seahorse is and swims more like a snake than a fish. It can also anchor itself to vegetation just like a seahorse by wrapping its tail around it. It usually hides in vegetation until a tiny animal swims near, and then it uses its tube-shaped mouth like a straw to suck in water along with the animal. Just like the seahorse, the male pipefish has a brooding pouch and takes care of the eggs after the female deposits them in his pouch.

So where did the rumor of seahorses in the Mekong come from? The Mekong is a river in southeast Asia that runs through at least six countries, including China, Thailand, Cambodia, and Vietnam. Parts of it are hard to navigate due to waterfalls and rapids, but it's used as a shipping route and there are lots of people who live along the river. Like all rivers, it's home to many interesting animals, but there's no evidence of seahorses anywhere throughout the Mekong's 2,700 mile length, or 4,350 kilometers.

But there is a hint about where the rumor of a Mekong seahorse could have come from. One researcher named Heiko Bleher chased down the type specimens of the supposed Mekong seahorse in a Paris museum, which were collected in the early 20th century by a man named Roule. Roule got them in Laos from a fisherman who had nailed the dried seahorses to his fishing hut. The fisherman told Roule the seahorses were from the Mekong, but when they were further studied in 1999 Roule's specimens were discovered to actually be specimens of *Hippocampus spinosissimus* and *Hippocampus barbouri*. Both are marine fish but do sometimes live in brackish water at the mouth of the Mekong. The fisherman wasn't mistaken, it's just that Roule misunderstood what he meant.

As for the freshwater seahorse supposedly found in Lake Titicaca, that one's less easy to explain. Titicaca is a freshwater lake in South America, specifically in the Andes Mountains on the border of Bolivia and Peru. It's the largest lake in South America and is far, far above the ocean's surface—12,507 feet above sea level, in fact, or 3,812 meters. It's also extremely deep,

932 feet deep in some areas, or 284 meters. It's home to many species of animal that live nowhere else in the world, including 90% of its fish. Why couldn't it be home to a freshwater seahorse too?

Titicaca was formed when a massive earthquake some 25 million years ago essentially shoved two mountains apart, leaving a gap—although technically it's two gaps connected with a narrow strait. Over the centuries rainwater, snowmelt, and streams gradually filled the gaps, and these days five rivers and many streams from higher in the mountains feed water into the lake. Water leaves the lake by the River Desaguadero and flows into two other lakes, but those lakes aren't connected to the sea. Sometimes they dry up completely. So Titicaca isn't connected to the ocean and never was, and even if it was, seahorses are weak swimmers and would never be able to venture up a river 12,000 feet above sea level. There's just simply no way a population of seahorses could have gotten into the lake in the first place, even if they could survive there.

That doesn't mean there aren't any freshwater seahorses ready to be discovered somewhere, of course. I just don't think you're going to find any in Lake Titicaca.

TRINITY ALPS
SALAMANDER

In the 1920s, an attorney named Frank L. Griffith, who was hunting in the area, spotted five salamanders in a lake in the Trinity Alps in northern California. They weren't ordinary salamanders, though. They ranged from 5 to 9 feet long, or 1.5 and 2.7 meters. He hooked one with a line, but he wasn't strong enough to land it and it escaped. In the 1940s, animal handler Vern Harden claimed he'd seen 8-foot, or 2.4-meter, salamanders in Hubbard Lake.

There are hundreds of salamander species throughout the world, most of them no more than a few inches long, or about 5 or 6 centimeters, but three are much bigger. The biggest is the Chinese giant salamander, which can grow almost 6 feet long, or 2 meters. The closely related Japanese giant salamander is almost as big, some 5 feet long, or 1.5 meters. A third giant salamander in the southeastern United States is smaller at 2.5 feet long, or 76 centimeters, called the hellbender. Larval hellbenders look a lot like another large salamander in the area, called the mudpuppy or water dog. The mudpuppy can grow a bit over a foot in length, or 31 centimeters, but it retains its gills throughout its life. Don't be fooled by fake hellbenders.

Giant salamanders are flattish in shape with broad bodies and wide heads. Their feet have stubby little toes. They eat fish, snails, crawdads, worms, insects, small mammals, snakes, frogs—basically anything they can catch. They catch their prey by opening their huge mouths quickly underwater, creating a vacuum that sucks small animals right in. They range in color from slate gray to black to brownish with dapples, but occasionally an orangish or pink individual is discovered.

All three species have thick folds of skin along their sides, which increases their surface area. That's important because they breathe through their skins. Larval giant salamanders have gills, but when they mature they lose those gills. The hellbender may retain a gill slit but it no longer functions. The salamanders need fast-moving water because it's well oxygenated. Giant salamanders are fully aquatic, although they can and do get out of the water occasionally for short periods. While they do have a single lung, they don't use it to breathe. They use it for buoyancy.

All the giant salamanders have poor eyesight, but they have a good sense of smell. In addition, the Chinese and Japanese giant salamanders have sensory cells along the sides of their bodies that detect vibrations in the water. The hellbender has light sensitive cells on its body instead, especially its tail. This lets it know when its tail is safely hidden, rather than sticking out from under a rock.

As far as we know, though, there are no giant salamanders in the Trinity Alps or anywhere else in the western United States. But the Trinity Alps do have the right climate with the right conditions for giant salamanders to thrive.

Thomas L. Rodgers, a biologist at Chico State College, conducted four expeditions to the Trinity Alps in 1948 in search of the giant salamander. The expeditions didn't find anything bigger than foot-long, or 30-centimeter, Pacific giant salamanders. In 1960, Bigfoot hunter Tom Slick convinced an expedition looking for Bigfoot to hunt for the salamander too, with no luck. Also in 1960, Tom Rodgers mounted another expedition, this time with some zoology professors and ten interested laymen. Again, they only found the Pacific giant salamander.

Rodgers decided he was wrong about the existence of a new giant salamander. In 1962 he denounced the previous sightings as misidentifications and hoaxes. More recently, a 1997 expedition led by Japanese-American writer Kyle Mizokami likewise came up with no sightings.

Herpetologist George S. Myers published a paper in 1951 about his own sighting. He said that in 1939 he was contacted by a commercial fisherman who had dredged up a 2.5-foot, or 76-centimeter, salamander in a catfish net from the Sacramento River. Myers described the salamander as dark brown with dull yellow spots, and said that it resembled the Chinese and Japanese giant salamanders but appeared to be a different species.

Tom Rodgers also saw the Sacramento specimen. The fisherman had managed to keep it alive in his bathtub. Rogers identified it as a Chinese giant salamander, and in fact it turned out to be a lost pet named Benny that had escaped while being taken to Stockton Harbor by steamer.

It's possible that giant salamanders once lived in the Trinity Alps but have since gone extinct. It's also possible they're still hiding out in the more remote parts of the mountain range. Then again, they might just be a quirky version of a fish story about the one that got away.

LAGARFLJÓT WORM

The Lagarfljót worm is a monster from Iceland, which is said to live in the lake that gives it its name. The lake is a pretty big one, 16 miles long, or 25 kilometers, and about a mile and a half wide at its widest, or 2.5 kilometers. It's 367 feet deep at its deepest spot, or 112 meters. It's fed by a river with the same name and by other rivers filled with runoff from glaciers, and the water is murky because it's full of silt.

Sightings of the monster go back centuries, with the first sighting generally thought to be from 1345. Iceland kept a sort of yearbook of important events for centuries, which is pretty neat, so we have a lot of information about events from the 14th century on. An entry in the year 1345 talks about the sighting of a strange thing in the water. The thing looked like small islands or humps, but each hump was widely separated, hundreds of feet from each other or at least 60 meters. The same event was recorded in later years too.

There have been strange sightings right up to the present day too. There's even a video taken of what surely does look like a slow-moving serpentine creature just under the water's surface.

The video was taken in February of 2012 by a farmer who lives in the area. Unlike a lot of monster videos it really does look like there's something swimming under the water. It looks like a slow-moving snake with a bulbous head, but it's not clear how big it is. A researcher in Finland analyzed the video frame by frame and determined that although the serpentine figure under the water looks like it's moving forward, it's actually not. The appearance of forward movement is an optical illusion, and the researcher suggested there was a fish net or rope caught under the water and coated with ice, which was being moved by the current.

But, of course, the video isn't the only evidence of something in the lake. If those widely spaced humps in the water aren't a monstrous lake serpent of some kind, what could they be?

One suggestion is that huge bubbles of methane occasionally rise from the lake's bottom and get trapped under the surface ice in winter. The methane pushes against the ice until it breaks through, and since methane refracts light differently from ordinary air, it's possible that it could cause an optical illusion from shore that makes it appear as though humps were rising out of the water. This actually fits with stories about the monster, which is supposed to spew poison and make the ground shake. Iceland is volcanically and geologically highly active, so earthquakes that cause poisonous methane to bubble up from below the lake are not uncommon.

Unfortunately, if something huge did once live in the lake, it would have died by now. In the early 2000s, several rivers in the area were dammed to produce hydroelectricity, and two glacial rivers were diverted to run into the lake. This initially made the lake deeper than it used to be, but has also increased how silty the water is. As a result, not as much light can penetrate deep into the water, which means not as many plants can live in the water, which means not as many small animals can survive by eating the plants, which means larger animals like fish don't have enough small animals to eat. The ecosystem in the lake is starting to collapse. Some conservationists warn that the lake will silt up entirely within a century at the rate sand and dirt are being carried into it by the diverted rivers.

At the moment, though, the lake does look beautiful on the surface. You probably won't see the Lagarfljót worm, but you never know.

~

Inkanyamba

THE INKANYAMBA IS a mystery water monster from South Africa. It's supposed to be some 20 feet long, or 6 meters. It lives in lakes and near waterfalls and is generally supposed to look like a snake or eel with a horselike head.

The inkanyamba seems to be associated with storms and other severe weather, an association that goes back untold centuries to cave paintings of what are known as rain animals. Groups such as the Xhosa and the Zulu believe that Inkanyamba is a giant winged snake that appears as a tornado as he flies around looking for his mate, who lives in a lake. Houses with metal roofs that aren't painted are in danger from Inkanyamba since he might mistake the roof for water.

Then again, there are sightings. In 1962 a park ranger saw an eel-like or snake-like creature on a sand bank along the Umgeni River, which slithered into the water as he approached. Another witness sighted the monster twice near Howick Falls in 1971 and 1981. He said it was 30 feet long, or 9 meters, with a crest along its neck. The waterfall known in English as Howick Falls in South Africa is sacred to the Zulu, who believe it's the home of Inkanyamba. It's 310 feet high, or 95 meters, and is situated on the Umgeni River. The only people who are traditionally allowed to approach the pool at the base of the falls, or who can safely approach it, are sangomas, or traditional healers.

One suggestion is that the inkanyamba is a giant mottled eel, which has fins that run all around the tail like a crest. But it only grows to about 6.5 feet long at most, or 2 meters. It's nocturnal, spends most of its time at the bottom of the lakebed or riverbed, and migrates from freshwater into the ocean to spawn and lay eggs.

The inkanyamba seems to be like the thunderbird, a creature of spiritual belief rather than a physical one. If you're lucky enough to visit Howick Falls, don't get too close to the water, out of respect for a sacred place...and just in case there's something there that could eat you up.

YEMISH

The iemisch, or hyminche, or lemisch, or some other variation, is often called a water tiger but linked not with a feline at all, but with a ground sloth. This is entirely the fault of a single man, Florentino Ameghino.

Ameghino was from Argentina, born to Italian immigrants, and is still highly regarded as a paleontologist, anthropologist, zoologist, and naturalist, from back in the days when you could specialize in lots of disciplines and still do tons of field work. He has an actual crater on the moon named after him. You don't get a moon crater unless you're pretty awesome. But Ameghino had at least one bee in his bonnet, and it involved giant ground sloths like megatherium. He was convinced they were still alive in the remote areas of South America, especially Patagonia.

In an 1898 paper he wrote about the yemish in Patagonia, which he said was so tough its hide couldn't be hurt by arrows or even fire. He said his brother Carlos, who was also a paleontologist, had sent him a piece of hide reputedly from a yemish, which he had gotten from a Tehuelche hunter. The hide had tiny bones embedded in it, called osteoderms, which are a feature of giant ground sloths. Ameghino claimed that the yemish was a giant ground sloth, which he named Neomylodon.

Mylodon, as opposed to Ameghino's Neomylodon, was a 10 foot long, or

3 meter, ground sloth that did indeed have osteoderms embedded in its thick hide. It had long, sharp claws and ate plants, probably dug huge burrows, and lived throughout Patagonia and other parts of South America. The important thing here is that mylodon remains, including dung as well as the dead animals, have been found in caves in Patagonia, and the remains look so fresh that the discoverers thought they were only a few years old. It turns out that they're all about 10,000 years old but were preserved by the cold, dry conditions in the caves.

So the piece of hide was probably really from a giant ground sloth, but not one that had been alive recently. Most researchers think that the sloths of Patagonia were already extinct when the area was first settled by humans, but discoveries of what looked like recently dead animals with fearsome claws and a hide that couldn't be pierced with arrows might very well have contributed to stories of local monsters.

But that's beside the point, because once you get past Ameghino's obsession with the yemish being a real live giant ground sloth, it's clear it's something completely un-slothlike. The exact term yemish isn't known from any language in Patagonia, but it might be a corruption of hymché, a water monster, or yem'chen, which means water tiger in the Aonikenk language. An even closer match from the same language means sea wolf and is pronounced ee-m'cheen [iü'mchün]. Other languages in the area call the elephant seal yabich, which also sounds similar to yemish. In other words, it's pretty clear that the yemish is a water animal of some sort.

The sea wolf is what we call a sea lion, a type of huge seal. Sea lions and elephant seals sometimes come up rivers and into freshwater lakes, which may account for some of the numerous lake monster legends in Patagonia. As for the hymché, it may have a natural explanation too that is nevertheless just as mysterious as just calling it a monster.

French naturalist André Tournouer explored Patagonia in 1900, and at one point while following a stream, he and his expedition saw what their guide called a hymché. It was the size of a large puma but with dark fur, rounded head, no visible ears, and pale hair around the eyes. It sank under the water when Tournouer shot at it, and later they found some catlike tracks in the sand along the bank.

From the description, it's possible that the hymché was a spectacled bear. It lives in the Andes Mountains of South America but was formerly much more widespread, and is usually black with lighter markings around the eyes that give it its name. Its ears are small and its head is more rounded than other bears. While it spends most of its time in the treetops, it actually does swim quite well. But as far as we know, spectacled bears don't live in Patagonia.

So, back to the yemish. According to Ameghino's 1898 paper, he said the Tehuelche referred to it as the water tiger. Since there is no local word for tiger in South America, because tigers live in Asia, this is probably a translation of the local word for puma. The jaguar did formerly live in Patagonia but was hunted to extinction there over a century ago.

The yemish supposedly spent much of its time in the river, where it dragged horses and other animals into the water when they came to drink. Its feet were flat, its ears tiny, it had big claws and fangs, and its toes were webbed for swimming. It had shorter legs than a puma but was bigger than one.

This sounds like one specific animal that does live in Patagonia, and it's not a tiger or any kind of feline at all. It may be an otter. Flat feet with claws and webbed toes? Check. Tiny ears and scary teeth? Check. Longer than a puma but with much shorter legs? Check. Otters don't kill animals as big as horses, of course, but they will scavenge on freshly dead animals. One story of a mule that fell off a precipice onto a river bank, and was discovered dead and half-eaten the next morning with strange paw prints all around it, fits with an otter family having an unexpected feast delivered to their doorstep.

Stories of monstrous otter-like animals are common throughout much of South America, not just Patagonia, and are frequently translated as "river tiger." In his book *Monsters of Patagonia*, Austin Whittall wonders why some tribes have two names for the otter in that case, an ordinary name and a name denoting a monster. It's possible the monster version of the otter either refers to a folkloric beast, an animal like a sea lion that was once seen far from its ordinary home, or two kinds of otter in the area, one bigger and more ferocious than the other.

The southern river otter lives in Patagonia, both in rivers and along the seashore. It's not especially big, maybe 4 feet long including the tail, or 1.2 meters. The rare marine otter also lives along the western and southern coasts of Patagonia. Its scientific name, *Lontra felina*, means "otter cat," and in Spanish it's often called *gato marino*, or sea cat. But the marine otter is relatively small, around 5 feet long at the very most, or 1.5 meters.

But there's another otter in South America, the giant otter. It lives north of Patagonia and is now endangered, with only around 5,000 animals left in the wild after being hunted extensively for its fur for decades. It's protected now, although loss of habitat and poaching are still big problems. It grows to around 6 feet long now, or 1.8 meters, but when it was more common some big males could grow over 8 feet long, or 2.5 meters. If in the past an occasional giant otter—twice the length of a southern river otter—strayed into the rivers of Patagonia, it would definitely be seen as a monster.

Whatever it is—real animal or a monster of folklore—one thing is clear. It's not any kind of sloth.

PART FOUR
SEA MONSTERS

I bet you flipped directly to this section, didn't you? Sea monsters are exciting because there's so much we don't know about the oceans. We'll learn about a few sea serpents in this section, along with some giant squid, deep-sea mystery fish, and a couple of mysterious whales.

BEEBE'S DEEP-SEA MYSTERY FISH

W illiam Beebe was an American naturalist born in 1877 who lived until 1962, which is amazing considering he made repeated dives into the deep sea in the very first bathysphere in the early 1930s. Even today descending into the deep sea is dangerous, and a hundred years ago it was way way *way* more dangerous.

Beebe was an early conservationist who urged other scientists to stop shooting so many animals. Back then if you wanted to study an animal, you just went out and killed as many of them as you could find. Beebe pointed out the obvious, that this was wasteful and didn't provide nearly as much information as careful observation of living animals in the wild. He also pioneered the study of ecosystems, how animals fit into their environment and interact with it and each other.

While Beebe mostly studied birds, he was also interested in underwater animals. Really, he seems to have been interested in everything. He studied birds all over the world, was a good taxidermist, and especially liked to study ocean life by dredging small animals up from the bottom and examining them. He survived a plane crash, was nearly killed by an erupting volcano he was observing, and fought in World War I. Once when he broke his leg during an expedition and had to remain immobilized, he had his bed carried outside every day so he could make observations of the local animals as they grew used to his presence.

In the 1920s, during an expedition to the Galapagos Islands, he started studying marine animals more closely. First he just dangled from a rope over the surface of the ocean, which was attached to a ship's boom, but eventually he tried using a diving helmet. This was so successful that he started thinking about building a vessel that could withstand the pressures of the deep sea.

With the help of engineer Otis Barton, the world's first bathysphere was invented and Barton and Beebe conducted dozens of descents in Bermuda, especially off the coast of Nonsuch Island. The bathysphere had two little windows and a single light that shone through one of the windows, illuminating the outside just enough to see fish and other animals. The bathysphere couldn't descend all that deeply, although it set records repeatedly. The deepest they descended was 3,028 feet, or 923 meters.

Beebe made careful notes of all the animals he observed from the bathysphere and published numerous articles and books about them. Many of these articles and books were illustrated by an artist named Else Bostelmann, who worked closely with Beebe and his team of scientists. Bostelmann even painted underwater while wearing a diving helmet, because she needed to know how colors were affected by underwater light. She used oil paints, since oil and water don't mix so the paints wouldn't wash away, and she tied strings to her paintbrushes so they wouldn't float off.

Many of the animals Beebe saw from the bathysphere have since been identified and described by later scientists. But there are five fish Beebe observed that have never been seen since.

Before we talk about them, let's learn about the Pacific blackdragon, for reasons that will shortly become clear. The Pacific blackdragon is a type of fish that lives in the Pacific, which you probably figured out without me telling you. It prefers tropical and temperate parts of the ocean, although since it's a deep-sea fish the water where it lives is mostly very cold.

The male Pacific blackdragon never eats. He *can't* eat. He doesn't have a functioning digestive system. He survives on the yolk from the egg he develops from and never grows any larger than his larval form, about 3 inches long, or 8 centimeters. He lives long enough to mate and then he dies.

The female, however, grows up to about 2 feet long, or 61 centimeters. Her body is long and thin, and her mouth is full of sharp teeth that she uses to grab anything she can catch. She especially likes to eat fish and small crustaceans, but she's not picky.

Her body is black, and not just regular black. It's called superblack or ultrablack, which absorbs most of the light that hits it. The Pacific blackdragon is superblack almost all over to help hide in the darkness of the water, since it's an ambush predator. Just under the fish's skin there's a layer of closely packed pigment-containing structures called melanosomes, which can absorb up to 99.95% of light. As if that wasn't enough, because a lot of the animals the blackdragon eats emit bioluminescent light, her stomach is also black to block any light from the prey she's swallowed. But although she's basically invisible to other animals, she does have several rows of light-emitting cells called photophores along her sides. Scientists think she uses the lights to attract a mate, but she only flashes the photophores occasionally and only for brief moments. She also has a barbel that hangs from her chin with a luminescent lure at the end, which she uses to attract prey.

While the Pacific blackdragon is a deep-sea fish, at night she migrates upward nearer the surface to catch more prey, although she still stays below about 1,300 feet deep, or 400 meters. She has large eyes as a result to take advantage of any moonlight and starlight that shines down that far. During the day she stays deeper, up to 3,200 feet deep, or 1,000 meters.

Speaking of the Pacific blackdragon's eyes, larval blackdragons have eyes on long stalks—really long stalks, nearly half their body length. As the larva matures, it absorbs the stalks until the adult fish has ordinary fish eyes. The larvae are also mostly transparent.

There are two other blackdragon species known, both of them a little smaller than the Pacific blackdragon. But in 1932 William Beebe spotted a fish that he thought might be related to the blackdragons, except that he estimated it was 6 feet long, or 1.8 meters.

Beebe named the fish *Bathysphaera intacta*, but there's no type specimen so no one can study it and verify whether it's a species of blackdragon or something else. Beebe said the fish he saw had large eyes, lots of teeth, and photophores along its sides that glowed blue, and had a barbel with a light under its chin just like the Pacific blackdragon and its cousins. But it

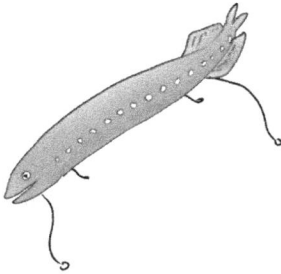

also had another, smaller barbel with a light near the tail. Beebe saw two of the fish together. They circled the bathysphere a few times, probably attracted to its light.

Another of Beebe's mystery fish is one he named the pallid sailfin, *Bathyembryx istiophasma*. He saw it twice on the same descent in 1934 and described it as about 2 feet long, or 61 centimeters, shaped like a cigar with triangular fins and a tiny tail. In fact, in his book *Half Mile Down* Beebe described the fish this way:

> The strange fish was at least two feet in length, wholly without lights or luminosity, with a small eye and good-sized mouth. Later, when it shifted a little backwards I saw a long, rather wide, but evidently filamentous pectoral fin. The two most unusual things were first, the color, which, in the light, was an unpleasant pale olivedrab, the hue of water-soaked flesh, an unhealthy buff. It was a color worthy of these black depths, like the sickly sprouts of plants in a cellar. Another strange thing was its almost tailless condition, the caudal fin being reduced to a tiny knob or button, while the vertical fins, taking its place, rose high above and stretched far beneath the body, these fins also being colorless.

Beebe assigned the pallid sailfin into the family Stomiidae, the same family that *Bathysphaera intacta* is assigned to as well as the other black-dragons. As a group, the fish in this family are called barbeled dragonfish. Some species in this family do show a similar tail arrangement that Beebe noted, with a very small tail fin but enlarged anal and dorsal fins that are set well back on the body

Another of Beebe's mystery fish was one he named the three-starred anglerfish, *Bathyceratias trilychnus*, which he estimated was about 6 inches long, or 15 centimeters. It had three bioluminescent illicia on its head that it probably used as lures, since that's something that other deep-sea anglerfish do and Beebe was pretty sure it was a species of anglerfish. Since there are over 200 known species of anglerfish, it's not surprising that there are more that aren't known.

Another was the five-lined constellation fish, *Bathysidus pentagrammus*, named for the five rows of photophores on its sides. Beebe thought it looked kind of like a surgeonfish, which is a flat, round fish shaped sort of like a pancake with fins and a tail. But surgeonfish are mostly found in shallow, tropical waters around coral reefs. They're often brightly colored. Beebe didn't assign his constellation fish to the surgeonfish's family, and in fact didn't assign it to any family since he didn't know where it belonged.

The last of Beebe's mystery fish was the abyssal rainbow gar, which he didn't give a scientific name to since he had no idea what kind of fish it might be. He thought it was shaped like a gar, but it was so extraordinary he didn't know what to think. He actually saw four of them swimming almost vertically, heads up and tails down, at about 2,500 feet deep, or 760 meters. He named them rainbow gar because of their coloring: bright red head and jaws, a light blue body, and a yellow tail. They were about 4 inches long, or a little over 10 centimeters, with long, pointed jaws. They moved by fanning the dorsal fin.

Beebe wrote scientific articles about some of these fish and included them all in his book *Half Mile Down*. But it wasn't long before other scientists started doubting the sightings. Some people thought he'd made up the fish to make his expeditions more exciting while some thought he was just mistaken. One irate ichthyologist wrote in 1933 that the constellation fish was probably just light reflecting off Beebe's own breath fogging the window, because no fish had photophores like the ones he described. I guess in 1933 everything was known about fish that would ever be known.

Beebe seems to have been an honest scientist, though, and he didn't need to make anything up. He discovered dozens, if not hundreds, of fish new to science, many of which have either been found and properly described later, or which Beebe himself managed to later catch. Whenever he and Barton came up from a descent in the bathysphere, Beebe had his team on the boat send down nets, and sometimes they caught some of the animals he had seen. This allowed Bostelmann to add details to her paintings that Beebe wouldn't have known about from just a look through the bathysphere's windows.

Not only that, if Beebe wanted to make up a fish that would excite the general public and make them want to buy his books, he would have made up something huge and frightening. His mystery fish are mostly quite small. Only *Bathysphaera intacta* was large, and he only said they were about 6 feet long. That's big for a deep-sea fish, but remember that the bathysphere never made it to the really crushing depths of the abyss. It descended into the mesopelagic zone, which is extremely dark but not completely lightless. There's also a lot of life in this zone, and many fish that spend the day here migrate nearer the surface at night where they can find more food while still remaining hidden. The long-snouted lancetfish lives in this zone and it can grow 7 feet long, or 2.15 meters.

Plus, Beebe didn't need to convince anyone to buy his books. They were already runaway bestsellers and he was quite famous, although it seems not to have gone to his head. He just wanted to have fun and do science. He actually seems to have been a good person by modern standards too, which is always refreshing. He disagreed with people who claimed to have scientific proof that women were inferior to men or that some races were inferior to others. He insisted that his team members work hard but he worked hard too, and if he thought everyone was feeling too stressed, he'd announce that his birthday was coming up and they should take a few days off to celebrate. Some years he had several birthdays.

Beebe did spot one other mystery animal, but he didn't get a good enough view to make a guess as to what it might be. This is what he wrote about it:

I saw its complete, shadow-like contour as it passed through the farthest end of the beam [of light]. Twenty feet is the least possible estimate I can give to its full length, and it was deep in proportion. The whole fish was monochrome, and I could not see even an eye or a fin. For the majority of the 'size-conscious' human race this marine monster would, I suppose, be the supreme sight of the expedition. In shape it was a deep oval, it swam without evident effort, and it did not return. That is all I can contribute, and while its unusual size so excited me that for several hundred feet I kept keenly on the lookout for hints of the same or other large fish, I soon forgot it in the (very literal) light of smaller, but more distinct and interesting organisms. What this great creature was I cannot say. A first, and most reasonable guess would be a small whale or blackfish. ...[O]r, less likely, it may have been a whale shark, which is known to reach a length of forty feet. Whatever it was, it appeared and vanished so unexpectedly and showed so dimly that it was quite unidentifiable except as a large, living creature.[1]

Twenty feet is 6 meters, by the way. It might easily have been a whale, since many species of whale routinely dive much farther than the bathysphere descended at its deepest. Whatever it was, and whatever Beebe's other five mystery fish were, hopefully one day a modern deep-sea vehicle will find them again.

LUSCA

T he lusca is an octopus-like sea monster from Caribbean folklore. The Caribbean Sea is part of the Atlantic Ocean outside of the Gulf of Mexico, and within the Caribbean Sea are thousands of islands, some tiny, some large, including those known collectively as the West Indies. Many reports of the lusca come from the Bahamas, where so-called blue holes dot many of the islands.

Blue holes are big round sinkholes that connect to the ocean through underground passages. Usually the holes contain seawater but some may have a layer of fresh water on top. The islands of the Bahamas aren't the only places where blue holes exist by any means. Australia, China, and Egypt all have famous blue holes.

Blue holes form in land that contains a lot of limestone. Limestone weathers more easily than other types of rock, and most caves are formed by water percolating through limestone and slowly wearing passages through it. This is how blue holes formed too. During the Pleistocene, when the oceans were substantially lower since so much water was locked up in glaciers, blue holes formed on land and many of them were later submerged when the sea levels rose. They can be large at the surface, but divers who try to descend into a blue hole soon discover that it pinches closed and turns into twisty passages that eventually reach the ocean, although no diver has been able to navigate so far. Many, many divers have died exploring blue holes.

Andros Island in the Bahamas has 178 blue holes on land and more than 50 in the ocean surrounding the island. It's also the source of a lot of lusca reports.

So what does the lusca look like? Reports describe a monster that's shark-like in the front with long octopus-like legs. It's supposed to be huge, with an armspan of 75 feet, or 23 meters, or even more. The story goes that the tides that rise and fall in the blue holes aren't due to tides at all but to the lusca breathing in and out.

People really do occasionally see what they think is a lusca, and sometimes people swimming in a blue hole are dragged under and never seen again. Since blue holes don't contain currents, people assume it must be an animal living in the water that occasionally grabs a swimmer.

The problem is, there's very little oxygen in the water deep within a blue hole. Fish and other animals live near the surface, but only bacteria that can thrive in low-oxygen environments live deeper. So even though the blue holes are connected to the ocean, it's not a passage that most animals could survive. Larger animals wouldn't be able to squeeze through the narrow openings in the rock anyway.

But maybe they don't need to. Most blue holes have side passages carved out by freshwater streams flowing into the marine water, which causes a chemical reaction that speeds the dissolving of limestone. Some blue holes on Andros Island have side passages that extend a couple of miles, or several kilometers. It's possible that some of these side passages also connect to the ocean, and some of them may connect to other blue holes. Most of the blue holes and side passages aren't mapped since it's so hard to get equipment through them.

Still, as far as we know, there is no monster that looks like a shark with octopus-like legs. That has to be a story to scare people, right? Maybe not.

The largest octopus known to science is the giant Pacific octopus. The largest ever measured had an armspan of 30 feet, or about 9 meters, although since octopus arms are kind of stretchy and only dead ones can be reliably measured, most individuals appear much smaller in the wild. It lives in deep water and like all octopuses, it can squeeze its boneless body through quite small openings. When it swims, its arms trail behind it something like a squid's and it moves headfirst through the water. A big octopus has a big mantle with openings on both sides for the gills and an aperture above the siphon. The head of the octopus could easily be mistaken for the nose of a shark, with a glimpse of the openings assumed to be its partially open mouth. A large octopus could easily grab a human swimming in a blue hole and drag it to its side passage lair to eat. Big octopuses even eat sharks.

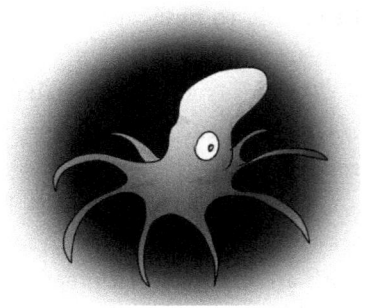

The giant Pacific octopus lives in the Pacific, though, not the Atlantic. If the lusca is a huge octopus, it's probably a species unknown to science, possibly one whose mantle is more pointy in shape, more like a squid's head. That would make it resemble a shark's snout even more.

Be careful where you swim.

GIANT MARINE INVERTEBRATES

Some of the largest animals on the planet are marine invertebrates. Scientists discover new ones every year, some of them of astonishing sizes. For example, in April of 2020, a deep-sea expedition off the coast of western Australia spotted several dozen animals new to science, including what may be the longest organism ever recorded. It's a type of siphonophore, which isn't precisely a single animal in the way that, say, a blue whale is. It's a colony of tiny animals, called zooids, all clones although they perform different functions so the whole colony can thrive. Some zooids help the colony swim while others have tiny tentacles that grab prey, and others digest the food and disperse nutrients to the zooids around them. Many siphonophores emit bioluminescent light to attract prey.

Some siphonophores are small but some can grow quite large. The Portuguese man o'war, which looks like a floating jellyfish, is actually a type of siphonophore. Its stinging tentacles can be 100 feet long, or 30 meters. Other siphonophores are long, transparent, gelatinous strings that float through the depths of the sea, snagging tiny animals with their tiny tentacles, and that's the kind this newly discovered siphonophore is.

The new siphonophore was spotted at a depth of about 2,000 feet, or 625 meters, and was floating in a spiral shape. The scientists estimated that the spiral was about 49 feet in diameter, or 15 meters, and that the outer ring alone was probably 154 feet long, or 47 meters. The entire organism might have measured 390 feet long, or almost 119 meters. It's been placed into the genus Apolemia although it hasn't been formally described yet.

Siphonophores are closely related to jellyfish, which aren't fish at all, of course. The largest species of jellyfish is probably the lion's mane jelly, although Nomura's jelly is about the same size. The lion's mane jelly lives in cold water and has a lobed bell, unlike most other jellies.

The lion's mane gets its name from the mass of over a thousand hair-like tentacles that somewhat resembles a lion's mane. Most of the tentacles aren't much longer than the bell diameter, but the outer tentacles can be extremely long. Each one is sticky and lined with stinging cells, as is the bell itself. An individual sting isn't very dangerous to humans, but there are so many tentacles with so many stinging cells that if a person gets tangled in them, the number of stings received can cause a dangerous reaction. The stings are also extremely painful.

The largest jellyfish ever measured was a lion's mane found off the coast of Massachusetts in 1870. It had a bell diameter of 7 feet, or over 2 meters, with tentacles 120 feet long, or 36.5 meters. This is huge, but there are reports of even bigger jellies sighted occasionally.

If you do a search online about giant jellyfish, you'll find an account of one much bigger than 7 feet across. The story goes that in 1973, an Australian ship called the *Kuranda* was nearing Fiji in the South Pacific when it collided with a gigantic jellyfish. The jelly didn't fare so well in the collision, more or less exploding on impact, but it was so huge that the goo from its bell covered the deck to a depth of about 2 feet, or 61 centimeters. The ship's captain, Langley Smith, estimated that the jelly's tentacles were more than 200 feet long, or 61 meters, and so many of them ended up on deck that one of the sailors got tangled in them and actually died from the multiple stings.

Even worse, the weight of the enormous jelly started to sink the ship. The *Kuranda* radioed an SOS and a salvage tug called the *Hercules* came to the rescue. It used a firehose to wash the massive amounts of goo and tentacles off the ship, saving her. Samples of the jelly were examined later in Sydney, Australia and were determined to be from a lion's mane jellyfish.

That's an amazing story—but it's not true. A librarian at the State Library Victoria in Australia helped me verify that there wasn't a ship named

Kuranda active in 1973, there wasn't a man named Langley Smith who was a ship's captain, and the Melbourne archive referenced in the original story (published in 1977) wasn't one that existed in 1973.

That doesn't mean there aren't stupendously large jellyfish in the oceans. Other ships have had trouble with jellies. In 2009, so many Nomura's jellyfish got tangled in a fishing net that it capsized the boat that was trying to haul the net in. This wasn't a little fishing boat, either; it was a ten-ton trawler. Fortunately the crew were all rescued.

Other marine invertebrates that are supposed to sink ships are gigantic octopus or squid. Squids in general have a body called a mantle, with fins at the rear and big eyes near the base. A squid has eight arms and two tentacles. The arms are lined with suction cups that contain rings of serrated chitin, which allows the squid to hang on to its prey. The tentacles usually only have suction cups at their ends. In the middle of the ring of arms, at the base of the mantle, is the squid's mouth, which is shaped like a parrot beak. Instead of actual teeth, the squid has a radula, which is basically a tongue studded with chitin that it uses to shred its prey into pieces small enough to swallow.

The giant squid is a deep-sea species that can grow an estimated 43 feet long, or 13 meters. Most of its length comes from its tentacles, though. The longest known giant squid's mantle and arms together reach around 16 feet long, or 5 meters, if you don't include the tentacles. Females are typically much bigger than males and can weigh twice as much.

The colossal squid is an even larger species, although it's not closely related to the giant squid. Some researchers estimate the colossal squid can grow to around 46 feet long, or 14 meters, but it has shorter tentacles and a much longer mantle than the giant squid, so is an overall much bigger and heavier animal.

The sizes of these two squid species are only estimates. We know very little about the colossal squid in particular. It was first described from parts of two arms found in the stomach of a sperm whale in 1925, and for more than 50 years that was pretty much all we had. Then a Russian trawler caught an immature specimen in 1981 off the coast of Antarctica. Since then researchers have been able to study a few other specimens caught or found dead.

As far as we know, the colossal squid is an ambush predator rather than an active hunter like the giant squid. It lives in the deep seas in the Southern Ocean, especially around Antarctica, as far down as 7,200 feet or 2.2 kilometers beneath the surface of the ocean, and it mostly eats fish. While its tentacles are much shorter than the giant squid's, they have something the giant squid doesn't. Its suckers have hooks, some of them triple-pointed and some of which swivel. When it grabs onto another animal, it's not letting go until somebody gets eaten.

All squid have fins of some kind on the mantle to help it move around. Different species, naturally, have varying sizes and shapes of fins. In the bigfin squid, as you may have already guessed, the fins are very big. They look more like wings and can be almost as large as the entire mantle. But that's not the really weird thing about these squid, although it was the most obvious thing when all we knew about them were young specimens. The arms and tentacles of bigfin squid don't develop to their full length until the squid is an adult.

In 2001, a deep-sea rover used by an oil company in the Gulf of Mexico caught video of a large, unusual squid. One of the men operating the rover remotely asked for a copy of the squid video for his girlfriend, who was interested in deep-sea animals. His girlfriend asked around, trying to find out what kind of squid it was, and eventually contacted a squid expert at the Smithsonian National Museum of Natural History. The squid expert is named Mike Vecchione and when he saw the video, he says he jumped out of his chair and started yelling in excitement. He'd never seen anything like this squid before.

Once he calmed down, he contacted his squid expert colleagues, who also freaked out. Eventually they found more footage of the weird squid taken by other oil rig rovers. The workers operating the rovers had no idea that the squid was a scientific mystery so hadn't thought to contact any scientists. Finally the squid was identified as an adult bigfin—the first adult ever seen by scientists.

In 2015, a deep-sea rover in a scientific expedition caught video of two bigfin squid near Australia, and in 2017 it saw three more. The rover was able to use lasers to get a much more accurate estimate of the squid's size than ever before. All five were different sizes so they were probably five different individuals. Measurements and size estimates of various bigfin squid suggest it can grow up to 26 feet long, or 8 meters, and maybe even longer.

The bigfin squid has very thin arms and tentacles, referred to as vermiform. That means worm-shaped, which gives you an idea of how thin we're talking. Unlike the octopus, which has eight arms (sometimes referred to as legs, and sometimes incorrectly referred to as tentacles), a squid has eight arms and two feeding tentacles.

In the bigfin squid, the arms and tentacles are the same size. In other squids, the tentacles are usually longer and look different from the arms. The great length of the arms and tentacles of the bigfin squid comes from what's called a distal filament that grows from the tip of each arm or tentacle. The filaments are sometimes missing, so it's probable that they're sometimes damaged and lost or maybe bitten off. If you don't count the distal filaments, the arms and tentacles are not actually all that long in comparison to the squid's mantle.

The bigfin squid can retract its filaments by coiling them up, but when it's hunting, it holds its arms and tentacles out from its body with the extremely long filaments hanging down. The filaments are sticky and trap tiny animals and particles of food drifting in the water.

A research sub investigating a WWII shipwreck spotted a bigfin squid 3.7 miles below the surface, or 6,000 meters, which makes it the deepest squid ever recorded. Imagine looking out the window of a submarine, trying to see details of a shipwreck, and suddenly there's a massive squid with incredibly long, thin arms looking back at you.

AKKOROKAMUI

A monster from Japan, Akkorokamui, has its origins in the folklore of the Ainu. The Ainu are a minority group of people who in the past mostly lived on Hokkaido, the second largest island in the country, although these days they live throughout Japan.

The story goes that a monster lives off the coast of Hokkaido, an octopus-like animal that in some stories is said to be 400 feet long, or over 120 meters. It's supposed to swallow boats and whales whole. But Akkorokamui isn't just an octopus. It has human features as well and godlike powers of healing. It's also red and because it's so big, when it rises near the surface of the water, the water and even the sky look red too.

Akkorokamui is supposed to be from the land originally. A humongous red spider lived in the mountains, but one day it came down from the mountains and attacked a town, stomping down buildings as the earth shook. The villagers prayed for help and the god of the sea heard them. He pulled the giant spider into the water, where it turned into a giant octopus.

The problem with folktales, of course, is that they're not usually meant to be taken at face value. Stories impart many different kinds of information, especially in societies where writing isn't known or isn't known by everyone. Folktales can give warnings, record historical events, and entertain listeners, all at once. It's possible the story of Akkorokamui is this kind of story, possibly one imparting historic information about an earthquake or tsunami that brought down a mountain and destroyed a town. That's just a guess, though, since I don't understand Japanese—and even if I did, the Ainu people were historically treated as inferior by the Japanese and many of their stories were never recorded properly. Much of their unique culture has been lost.

We don't know if Akkorokamui was once thought of as a real living animal, a spiritual entity, or just a story. There are a few reported sightings of the monster, but they're all old and light on details. One account from the 19th century is supposedly from a Japanese fisherman who saw a monster with tentacles as big around as a grown man. It was so big that the fisherman at first thought he was just seeing reflected sunset light on the ocean. Then he came closer and realized what he was looking at—and that it was looking back at him from one enormous eye. He estimated it was something like 260 feet long, or 80 meters. Fortunately, instead of swallowing his boat, the monster sank back into the ocean.

LONELIEST WHALE

You may have heard of the loneliest whale. The story goes that this whale is lonely because its voice is too high to be heard or understood by other whales. It calls but never gets a response.

That's actually not the case. Its voice is higher than other blue whales, fin whales, and humpback whales, but they can certainly hear it. For all we know, they answer. Since the individual whale hasn't actually been spotted, we don't know if it travels alone or with other whales.

The loneliest whale was first detected in 1989 by the U.S. Navy listening for submarines in the North Pacific, and again in 1990 and 1991. At that time the recordings were classified, but some were partially declassified in 1992 and word about the whale got out.

The calls vary but are similar to blue whale calls. The main difference is the voice's pitch. The loneliest whale calls at 52 hertz. That's slightly higher than the lowest notes on a piano or tuba. Blue whale songs are typically around 10 to 40 hertz. The whale's voice has deepened over the years to around 46 hertz, suggesting that it has matured.

Suggestions as to why this whale has a different call include the possibility that the whale is deaf, that it's malformed in some way, or that it's a hybrid of two different species of whale. Its migratory patterns suggest it's a blue whale. Its call patterns, if not its actual voice, resemble those of a blue whale but with some fin whale elements. Fin whales and blue whales do interbreed occasionally, but no one has successfully recorded a hybrid's calls.

Whale researchers think the recordings seem to be of one individual whale, but in 2010 sensors off the coast of California picked up lonely-whale-like calls that might have been made by more than one whale at the same time. One suggestion is that blue and fin whale hybrids might be common enough that they band together.

Ocean noise pollution from cargo ships, military operations, and other human-caused factors is a big problem for whales and other marine life. Some researchers suggest the whale's higher voice may be its attempt to cut through the noise and be heard by other whales.

≈

Scrag Whale

THE STORY GOES THAT SOMETIME BEFORE the year 1672, on Nantucket Island in the northeastern United States (although it was still a British colony back then), a scrag whale entered the harbor and hung around for three days. The white settlers wanted to kill it but they didn't have a weapon big enough. Finally the local blacksmith made a harpoon and the people killed the whale, which started the whaling industry that Nantucket became famous for.

But what's a scrag whale?

For a long time the accepted answer was that it was a young Northern right whale. Now the scrag whale is supposedly just a gray whale, young or not. But the gray whale only lives in the Pacific Ocean. Nantucket Island is in the north Atlantic.

The scrag whale was described as looking like a fin whale, but instead of a dorsal fin, it had bumps along its back. The gray whale doesn't really look that much like the fin whale, though, except that they're both baleen whales. The fin whale is much longer and more slender, and is actually a paler gray than the gray whale is.

The gray whale can grow almost 50 feet long, or about 15 meters. It's dark gray with grayish-white scars on its body. The whale migrates long distances from warm breeding grounds to cold feeding grounds, and its skin parasites die and drop off when the whales reach cold water, leaving scars behind. In fact, the gray whale migrates the farthest distance of any mammal known, a round trip of about 13,700 miles, or 22,000 kilometers.

The gray whale has short baleen plates, a pair of blowholes, and bumps called knuckles along the rear of its body, known as a dorsal ridge, that takes the place of a dorsal fin. It's not actually that closely related to other whales, although it's probably most closely related to humpbacks and fin whales. It's the only living member of its own family.

The only other member of the gray whale's family (that we know of, anyway) has been extinct for almost two million years, the Akishima whale, and we only have the fossilized remains of one animal. It grew 39 feet long, or 12 meters, and was found in a riverbed in Tokyo, Japan, by a man and his son. Fortunately, it's a beautifully preserved specimen. It was described in 2017.

But let's get back to the scrag whale. "Scrag" used to be a term used for whales that were in poor condition or that lived near the coast. If you've ever referred to a person or animal as scraggy or scraggly, meaning thin and dirty, you'll understand the term scrag whale even better. It's possible that the scrag whale wasn't actually a specific type of whale after all, but just a general term for any sad-looking whale.

Then again, it might have been a term for the gray whale in particular but any kind of sad-looking whale in general. Think about the term ratty. I call all kinds of things ratty, from my own hair after I tried to cut it myself, to a bird visiting my feeders in the rain with its feathers all wet and disordered. It doesn't mean I think my hair or the bird have anything to do with a rat.

Anyway, there are two living populations of gray whale, one in the eastern North Pacific and one in the western North Pacific, both endangered. But there used to be an Atlantic population too. The North Atlantic gray whale isn't well known, but we know it migrated from the eastern coast of North America to the western coast of Europe, and even into the Mediterranean Ocean—but the most recent remains found have been dated to about 1675. Early whalers drove it to extinction.

As the eastern Pacific gray whale slowly becomes more numerous now that it's a protected species, some individuals have been showing up in the Atlantic. In the last decade or so, a few gray whales were spotted in the

North Atlantic after they crossed there from the Pacific by way of the Arctic Ocean. All the world's oceans are connected, of course, and going from one to the other is just a matter of avoiding those pesky continents in the way. A whale like the gray whale that can easily travel 75 miles a day, or 120 kilometers, has no trouble crossing entire oceans if it wants to.

Hopefully the gray whale will eventually repopulate the Atlantic. This time we *promise* not to kill any!

DAEDALUS SEA SERPENT

On August 6, 1848, about five o'clock in the afternoon, the captain and some of the crew of HMS *Daedalus* saw something really big in the water. The ship was sailing between the Cape of Good Hope and St. Helena on the way back to England from the East Indies. It was an overcast day with a fresh wind, nothing unusual. The midshipman noticed something in the water he couldn't identify and told the officer of the watch, who happened to be walking the deck at the time with the captain. Most of the crew was at supper.

An artist's illustration of the monster

This is what the captain, Peter M'Quhae, described in his report when the ship arrived at Plymouth a few months later.

[A]n enormous serpent, with head and shoulders kept about four feet constantly above the surface of the sea, and, as nearly as we could approximate...there was at the very least sixty feet of the animal à fleur d'eau [at the water's surface]... The diameter of the serpent was about fifteen or sixteen inches behind the head, which was, without any doubt, that of a snake; [...] its colour a dark brown, with yellowish white about the throat. It had no fins, but something like a mane of a horse, or rather a bunch of seaweed, washed about its back.

The original *Times* article also mentioned large jagged teeth in a jaw so large that a man could have stood up inside the mouth, but this seems to be an addition by the article's writer, not the captain or crew.

The officer of the watch, Lieutenant Edgar Drummond, also published an excerpt from his own journal about the sighting, which appeared in a journal called the *Zoologist* in December 1848.

[T]he appearance of its head, which, with the back fin, was the only portion of the animal visible, was long, pointed, and flattened at the top, perhaps ten feet in length, the upper jaw projecting consider-ably; the fin was perhaps twenty feet in the rear of the head, and visible occasionally; the captain also asserted that he saw the tail, or another fin about the same distance behind it; the upper part of the head and shoulders appeared of a dark brown colour, and beneath the under jaw a brownish white. It pursued a steady undeviating course, keeping its head horizontal with the surface of the water, and in rather a raised position, disappearing occasionally beneath a wave for a very brief interval, and not apparently for purposes of respiration. It was going at the rate of perhaps from twelve to four-teen miles an hour....

To translate some of this into metric, 10 feet is about 3 meters, 20 feet a little over 6 meters, and the speed of 13 miles per hour is almost 21 kilometers per hour.

A lot of people wrote to the *Times* to discuss the sighting and suggest solutions. One writer claimed the animal couldn't be a snake or eel, since a side to side undulating motion would have been obvious as the animal propelled itself with its tail. Another said it had to have been a snake but the undulations were only in the tail, which was below the water. Yet another suggested it was a monstrous seal or other pinniped. Captain M'Quhai took exception to that one and wrote back stressing that he was familiar with seals and this definitely had not been one. Other suggestions included a basking shark or some other unknown species of shark, a plesiosaur, or a giant piece of seaweed.

Other similar sightings are on record, including one from the very end of 1849 off the coast of Portugal. An officer on HMS *Plumper* reported seeing a monster almost identical to the *Daedalus* sighting.

A long black creature with a sharp head, moving slowly, I should think about two knots, through the water, in a north westerly direction, there being a fresh breeze at the time, and some sea on. I could not ascertain its exact length, but its back was about twenty feet if not more above water; and its head, as near as I could judge, from six to eight. [...] There was something on its back that appeared like a mane, and, as it moved through the water, kept washing about, but before I could examine it more closely, it was too far astern.

Illustrations of the Daedalus sea serpent, which M'Quhai approved, were published in the *Times*. But the original sketch made by Drummond in his journal the day he saw the animal gives us a much better idea of what it looked like and what it probably was. The sketch accompanying the Plumper sighting reinforces the solution. It's probable that both sightings, and probably many others, were of a sei whale skim feeding.

The Plumper *sighting*

The sei is a baleen whale that's generally considered the fourth largest whale, with some individuals growing almost 65 feet long, or nearly 20 meters. Females are larger than males. It lives all over the world although it likes deep water that isn't too cold or too hot. It's a mottled dark grey in color with a tall dorsal fin fairly far back on the animal's body. Its tail flukes aren't usually visible. Its rostrum, or beak, is pointed and short baleen plates hang down from it. The sei whale's baleen is unusually fine, with a curly white fringe that looks something like wool.

Unlike some whales, it doesn't dive very deeply or for very long, and it's usually relatively solitary. It spends a lot of its time at or near the surface, frequently skim feeding to capture krill and other tiny food. It does this by cruising along with its mouth open, often swimming on its side. It has throat pleats that allow its huge mouth to expand and hold incredible amounts of water. The whale closes its mouth and raises its gigantic tongue, forcing the water out through its baleen plates. Whatever krill and fish are caught by the baleen, the whale swallows.

A lot of baleen whales skim feed occasionally, but the sei is a skim feeding specialist. It has a narrow, pointed rostrum that often sticks out of the water as it skim-feeds, with pale baleen hanging down. This might easily look like a long snaky animal with a small head held out of the water, especially in poor viewing conditions when the people involved are convinced they're looking at a sea serpent. The sei whale is a fast swimmer too, easily able to cruise at the speeds described by the *Daedalus* and *Plumper* crews.

It's not a perfect match, of course. The sei whale's dorsal fin is pretty distinctive and if seen properly would have immediately told the crew they were looking at a whale. No one reported seeing anything that could be considered a whale's breath either, sometimes called a spout. Since whales exhale forcefully and almost empty their lungs when they do, the cloud of warm air expelled looks like steam and is a tell-tale sign of a whale. Whales also don't have hair on their rostrum that could wash around like a mane on a sea serpent's neck. So while it seems likely that the *Daedalus* and *Plumper* sightings were of sei or other baleen whales skim feeding, we can't know for sure.

GLOUCESTER SEA SERPENT

While the Gloucester sea serpent was first mentioned in a traveler's journal in 1638, it really came to prominence almost two centuries later. On August 6, 1817, two women said they'd seen a sea monster in the Cape Ann harbor. A fisherman said he'd seen it too, but neither the fisherman nor the women were believed. A 60-foot, or 18-meter, sea serpent in the harbor? Ridiculous!

Only a few days later, though, the monster started showing up in Gloucester Bay and attracted major attention—not because it was elusive, but because it was so commonly seen. Sailors, fishers, and even people on shore saw what was described as a huge serpent in the waters of Gloucester Bay, Massachusetts, in the northeastern United States. On one occasion more than two hundred people watched it for nearly four hours.

The creature's length was described as anywhere up to 150 feet long, or 46 meters, and many people said it had a horse-sized head. Some people described its head as being about the same shape as a horse's too, although with a shorter snout. The body was snake-like and about the thickness of a barrel.

Many people thought the sea monster had humps along the back, usually referred to as bunches or occasionally joints. Others said it undulated through the water in an up-and-down motion, which looked like humps. Others said it had no bunches or humps at all. Most people agreed that its back was dark brown.

One of the earlier witnesses, a man named Amos Story, watched the sea serpent from shore for an hour and a half. He was adamant that it had no bunches, that he only saw at most about 12 feet of its length at one time, or 3.6 meters, and that its head resembled that of a sea turtle. It was also fast, with Story claiming it covered a mile in only three minutes or so. That's about 20 miles per hour, or 32 kilometers per hour—an incredible speed for an animal in the water.

As it happens, the leatherback sea turtle has been recorded as swimming that fast, and it can grow over 7 feet long, or 2.2 meters, and possibly much longer. It lives throughout the world's oceans and is just as happy in cold waters as it is in tropical waters. In other words, it's possible Story actually saw a huge leatherback turtle, which would explain why it had a turtle-like head that it held above the surface of the water at least part of the time. This is something leatherback turtles do. Then again, the leatherback has distinctive ridges and serrations on its back that Story didn't mention.

So many people reported seeing the sea serpent that the Linnaean Society of New England decided it needed to investigate. The society had only formed a few years before, in 1814, to promote natural history. By 1822 it had disbanded, but in those eight years it accomplished quite a bit, including opening a small museum in Boston. Its most controversial endeavor was the sea serpent investigation.

Members of the Linnaean Society interviewed witnesses, making careful notes that were signed by the interviewees to indicate the details were accurate. These statements tell us a lot about what people saw, although it hasn't helped us determine what the sea serpent actually was.

For instance, Captain Solomon Allen saw the creature more than once and gave a clear description of it. It was at least 90 feet long, or 27.5 meters, with as many as fifty joints, or bunches. Its head was snake-like—specifically rattlesnake-like, presumably meaning it was wider at the back and had a narrower snout—but the size of a horse's head. It was dark brown, plain in color, and swam with an undulating side-to-side motion. It dived by sinking straight down, moved quickly, and sometimes seemed to play in the water by swimming in circles.

All this is great information, but it doesn't resemble any known animal.

It also doesn't necessarily resemble the other witness statements. Let's go over some of the more detailed sightings and see if we can come to some conclusions.

A man named William Foster reported bunches along the monster's length, although he also described them as rings. When the animal's head rose from the water, the first thing Foster saw was what he described as a prong or spear. It was about a foot long, or 30 centimeters, and tapered to a point. His interviewer asked if the spear might have been a tongue, but Foster didn't think so.

Three men on a schooner named the *Laura*, becalmed in the mouth of the harbor, witnessed the monster in late August. Sewall Toppan, master of the ship, reported that the monster's head was the size of a 10-gallon keg, which would have been about 18 inches tall, or 46 centimeters, and 16 inches in diameter, or 40 centimeters. He said its head was held about 6 inches out of the water, or 15 centimeters, and that he could see 10 or 15 feet of its length disappearing into the water, or 3 to 4.5 meters. He didn't see any kind of prong, but two of his sailors did.

One of the two sailors was Robert Bragg, who reported that the monster was swimming rapidly toward the ship with its head and about 15 feet of its body out of the water, or 4.5 meters. As it drew closer he saw its tongue, which he described as looking like a harpoon about 2 feet long, or 61 centimeters. He even reported that the animal raised its tongue almost straight up several times. He also said it was dark brown and smooth.

The third *Laura* witness, helmsman William Somerby, corroborated Bragg's details, including the animal's tongue, which he mentioned was light brown. As the monster passed within 40 feet of the ship, or 12 meters, Somerby even saw one of its eyes clearly. He said it was the size of an ox's eye and was completely dark brown or possibly black. He and Bragg both noted that the animal had a bunch above its eyes, presumably meaning a bump or knob of some kind.

All three men said that they were familiar with whales and the animal was not a whale.

August 14 was a warm day and the water was calm. A man named Matthew Gaffney, a ship's carpenter by trade but in his heart a monster hunter, borrowed a boat and took his brother and a friend with him to row. He also took a musket.

As the small boat approached cautiously, the monster was spiraling around in the water, as various people reported it doing on and off throughout the day. Gaffney waited until the boat was as close as it could safely approach without risking being capsized, then fired a shot at the monster's head.

He was a good marksman and was certain he hit the animal, which sank immediately below the surface and vanished. Worried that the wounded monster would be enraged once its initial shock wore off, Gaffney and all the other boats on the harbor took off for shore. But when the sea monster resurfaced some distance off, it was obviously unbothered by being shot at. It continued its apparently playful circling around in the harbor.

Several witnesses who saw the monster on August 14, before and after Gaffney's attempt to shoot it, gave statements. William H. Foster said it at first moved slowly, but then sped up and twisted and turned through the water. Sometimes its head would bend around toward its tail, and Foster specifically said that when that happened, parts of its body between the bunches would raise up as much as 8 inches out of the water, or 20 centimeters, showing that the animal was at least 40 feet long, or 12 meters.

Lonson Nash saw the sea serpent and reported that it moved quickly and left a long wake, and that while it swam underwater sometimes, it didn't seem to be very far under. He could track its progress underwater by the disturbance it made on the surface. He also saw it double around so that its head was sometimes near its tail, but he mentioned that when it was swimming forward, it appeared perfectly straight.

Later that day, a shipmaster named Epes Ellery saw the monster's head through a spyglass. He reported that it was flattened on top like a snake's and that its mouth resembled a snake's mouth—presumably meaning it had a thin lower jaw. He reported that its joints were the size of two-gallon kegs and rose about 6 inches above the surface, or 15 centimeters. He said the animal swam with a vertical motion, not a side-to-side motion.

An unnamed woman reported that the sea monster's bunches looked like gallon kegs tied in a line. Another man said he saw the creature's bunches at the surface as it lay still for a while, and that around 50 feet, or 15 meters, of its length was visible although he couldn't see its head or tail. Other witnesses that same day reported much the same thing.

Captain Elkanah Finney saw the sea monster from shore later in August, after his son reported seeing something strange in the harbor. Finney first thought it was a bunch of seaweed, but when he looked at it through his spyglass he realized it was an animal moving quickly through the water. He said it might have been 100 feet long, or 30 meters, with 30 or 40 bunches down its length. In fact, he said it looked like a string of buoys and that each bunch was about the size of a barrel.

There are lots of other reports, all of them similar to these. The sea monster, whatever it was, spent a lot of time in and around Gloucester Bay that summer and even returned the following two summers. People were obviously seeing *something*. The question is what.

Fanciful 1817 painting of the sea serpent

Let's look first at the sightings where the monster had a prong or that it stuck out a long, straight tongue. This sounds a lot like a narwhal. A narwhal can grow up to about 18 feet long, or 5.5 meters, and males, and some females, have a brown or brownish spiral tusk that can grow just over 10 feet long, or 3 meters. Many people think the narwhal's tusk is a horn that sticks up from its forehead, but it's actually an elongated tooth that grows through the upper lip. That would explain why some of the witnesses thought it was a tongue.

A young narwhal is black or dark brown, although it grows lighter throughout its life so that old narwhals are almost white. A young animal would also have a short tusk. A narwhal often swims with its head out of the water and a male will sometimes lift his tusk up and down in the air. He can do this easily because, unlike most whales, the narwhal's neck vertebrae aren't fused and can bend the head around.

Most importantly, the narwhal is an Arctic animal and isn't typically found as far south as Massachusetts, although it's certainly been seen in that part of the ocean on rare occasions. Its rareness, together with its odd appearance compared to other whales, might lead witnesses to think it wasn't a whale at all but some kind of monster.

That doesn't explain the bunches, though. The witnesses on the schooner *Laura* didn't report seeing any bunches on their sea monster (whose "tongue" reportedly looked like a harpoon), but William Foster's pronged sea monster did have bunches.

Some researchers have dismissed the bunches, or humps, as a string of narwhals or other small whales traveling in a line. That's definitely a possibility, but too many witnesses described the bunches as being always partially out of the water, not moving up and down. Not only that, the bunches were seen when the sea monster was lying quietly on the placid surface, not moving, often for long stretches.

Remember, though, that many witnesses described the bunches as resembling a line of buoys or kegs tied on a line. The animal often seemed to swim in circles until its head nearly touched its tail. William Foster reported that when it did this, its body between the bunches would rise several inches out of the water. Lonson Nash said when it was swimming forward, its body appeared perfectly straight.

Maybe witnesses weren't seeing a long serpentine animal with bumps along its back. Maybe they were seeing a string of kegs used as buoys to keep fishing nets afloat, that had become tangled around a small whale's tail.

Small kegs or large pieces of cork were sometimes used for this purpose at the time, including in Newfoundland and Norway. If a net tangled around a narwhal's tail, the animal might have become used to dragging its burden around until the net eventually rotted away and freed the whale. This is something that still happens to whales today with nets and other fishing gear, although these days the nets are all plastic and won't rot.

Narwhals mostly eat fish and squid, and often dive deeply to find food along the ocean floor. Our tangled narwhal chasing fish underwater might appear to be traveling in playful circles as the net dragged along behind and above it. Pulling all the buoys underwater would probably be difficult for the whale, which would explain why it mostly stayed near the surface.

It's not a perfect match, of course, but the tangled-narwhal hypothesis fits a lot of the details reported for the Gloucester sea serpent. Narwhals also often travel in small groups, so if the tangled narwhal was with a few friends, that would explain why not every witness saw the bunches.

As for the Linnaean Society of New England, their investigation of the sea monster was excellent for the time. They took the sightings seriously and tried to remain impartial, although the members did seem to start from an assumption that the animal was an actual serpent of some kind.

Unfortunately, they made one fatal blunder. In late September 1817, someone found and killed a snake 3.5 feet long, or a little over a meter, that had bunches all down its spine. It was found only a few miles from Gloucester Harbor. The Linnaean Society decided it had to be a baby sea serpent.

They said so loudly and even proposed a scientific name for the sea serpent. But it wasn't long before the "baby sea serpent" was identified as a common black snake. The body was dissected and the bunches turned out to be tumors from a diseased spine. The society's investigation became a joke. At least we still have the eyewitness accounts they gathered.

VALHALLA SEA MONSTER

O
n December 7, 1905, two naturalists spotted an animal they couldn't recognize off the coast of Brazil.

The pair were Michael Nicoll and Edmund Meade-Waldo, part of a research team on the *Valhalla*. The ship was about 15 miles, or 24 kilometers, from the mouth of the Parahiba River. At 10:15 am Nicoll spotted a fin above the water about 100 yards away, or 91 meters. The fin was roughly rectangular, close to 2 feet high, or 61 centimeters, and 6 feet long, or 1.8 meters, and dark brown with an edge Meade-Waldo described as crinkled.

Meade-Waldo was looking at the fin through his binoculars when a head and long neck emerged from the water in front of the fin. He estimated it as 7 or 8 feet high, or over 2 meters, with a brown, turtle-like head. The animal moved its neck from side to side so strenuously that it thrashed the water into foam. They watched it until it was out of sight as the ship sailed away, but early the next morning, around 2 am, three crew members spotted what they thought was the same animal swimming underwater.

Nicoll agreed that the fin stuck up above the water about 2 feet, but thought it was closer to 4 feet long, or 1.2 meters. He also mentioned that it sometimes disappeared beneath the water and had an almost rubbery appearance.

They published their report the following year in *The Proceedings of the Zoological Society*. No one could identify the animal they spotted.

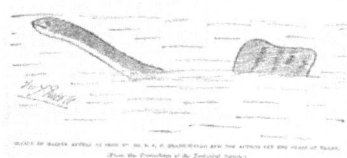

Nineteen years later, a British traveler named C.H. Prodgers published a book titled *Adventures in Peru*. He mentioned seeing a strange animal off the coast of Brazil in 1905—at about the same place that the naturalists saw their sea monster later the same year. He said its head was the size of a cow's head and its body was as large and round as a barrel. Instead of a fin, he described seeing a "coil" of the body sticking up about a foot out of the water, or 30 centimeters. He didn't mention what the head looked like but did say that the "coil" was 8 to 10 yards behind the head, or as much as 9 meters.

As we learned in the Gloucester sea serpent chapter, sea serpents were commonly thought to have odd structures along the back referred to variously as bunches, coils, joints, and so forth. Most likely these were usually floats from fishing nets and other fishing gear tangled around an animal's tail or body, stretched out behind it so that they looked like an enormously long knobby tail. That explains why Prodgers, who thought he'd seen a sea serpent, said he only saw one coil: he assumed there were more coils underwater.

That doesn't help us determine what the creature was, though. It had a relatively long neck and a turtle-like head, and a structure behind it that was presumably part of the body. Two witnesses thought it was a fin, another thought it was a barrel-shaped coil.

The solution might lie in the animal's behavior. The naturalists noted that the animal was thrashing its head and neck back and forth above the water. If it was a large marine turtle like a leatherback that had become entangled in a human-made object, it might have had to swim hard to drag its burden along, resulting in the thrashing motion. The rubbery fin might have actually been rubber.

The naturalists described the animal as having a turtle-like head and Prodgers mentioned that its body was as round as a barrel. The squared-off, rubbery fin with a crinkled edge might even have been a leatherback sea turtle's carapace seen at a distance through relatively primitive binoculars.

Instead of a regular hard shell like most other turtles, the leatherback's carapace is made up of leathery skin studded with osteoderms. It also has seven ridges that look like seams that run down the back from head to tail. These are raised and can appear jagged or crinkled. The leatherback often swims with its head out of the water, too.

It's not a perfect match, of course. Even the biggest leatherback doesn't grow to the sizes the witnesses described, although it can grow to about 9 feet long, or 2.75 meters. The fin might have actually been a fin of some unknown animal whose head resembled a turtle's.

OARFISH

If you think about a sea serpent, you probably imagine a big, slender animal that moves through the water like a snake even though it's not a snake. It may have a crest on its head or a mane of some kind. It's also really long.

Congratulations! You just described an oarfish, more or less.

The giant oarfish can definitely grow up to 36 feet long, or 11 meters, and possibly up to 56 feet long, or 17 meters. It may even sometimes grow longer than that. It's silvery in color with a red crest on its head and a mane-like fin down its spine, although it's actually an elongated dorsal fin. It has extremely long pelvic fins too.

We know so little about the oarfish, and what we know is so strange, that it's the next best thing to a sea serpent. The first living oarfish was only filmed in 2001. Most oarfish are only seen when they're dead or dying. It seems to live throughout the world's oceans, except for the Arctic and Antarctic, and is a deep-sea fish but may migrate closer to the surface at night to find more food.

The giant oarfish has a short, blunt snout and no teeth because it filters krill and other tiny animals from the water. But although it's long and slender like an eel, it actually swims vertically with its head pointing up and its tail down. We're not sure why, although one theory is that this minimizes its profile to predators looking up from below.

The oarfish doesn't have scales. Instead, its skin is soft with a delicate layer called ganoine that gives it a shimmery, almost metallic appearance. The long filaments of the crest on its head and its pelvic fins are also delicate, and its long tail often shows damage from being bitten. Since its organs are all close to the front of its body, and it doesn't need its tail for swimming, if a predator takes a bite out of its tail, the fish is going to be fine. It can swim quickly straight up and down to avoid predators that mostly just swim forward.

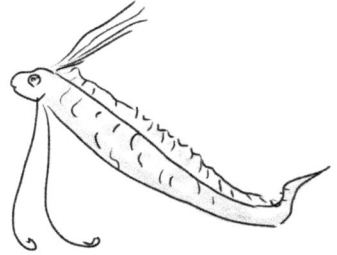

A Japanese legend says that another species of oarfish, the slender oarfish, predicts earthquakes. If an oarfish is seen near the surface or washes up on a beach, an earthquake is supposed to be imminent. That seems to be coincidence, though.

At least one species of oarfish, the streamer fish, is reportedly able to produce a weak electric shock. It's possible that the giant oarfish can too. The streamer fish can grow up to 10 feet long, or 3 meters, although since only seven specimens have ever been found, that might be a low estimate. It's only been found in the southeastern Pacific Ocean in both deep and shallow water and its skin is studded with rows of tiny hard nodules.

The oarfish looks like a sea serpent. Maybe that means it really is a sea serpent. It's certainly mysterious.

HOOK ISLAND SEA MONSTER

Hook Island is a small island off the coast of Australia, specifically Queensland in northeastern Australia. It's one of 74 islands in the area and hardly anyone lives on this particular island because most of it is a national park. Tourists visit it to dive among the coral off its northern coast and occasionally die from jellyfish stings. But for the most part it's just a quiet, picturesque part of the world.

But in 1964, a man named Robert Le Serrec was boating in the area with his family and a friend and took three pictures of something enormous in the water. It's known as the Hook Island Sea Monster.

Le Serrec was a French photographer, and in 1965 he wrote about his encounter with the sea monster in a magazine called *Everyone*. He, his wife and kids, and an Australian friend named Henk de Jong were spending the summer on Hook Island, which back then wasn't off-limits to foot traffic and in fact even had an inexpensive wilderness resort for people to stay in while they were there. Le Serrec had a motorboat and on December 12, 1964, which is summer in the southern hemisphere, the group was crossing Stonehaven Bay when Le Serrec's wife saw something at the bottom of the shallow water that looked like an enormous tadpole. And I mean *enormous*.

At first they estimated it was 30 feet long, or over 9 meters. But when Le Serrec and de Jong dived into the water to try and get film footage of it, they realized it was over twice that length—maybe as long as 90 feet, or over 27 meters. When it opened its mouth and started moving toward them, they decided to get out of the water again, and fast. While they were returning to the boat, though, the monster swam off.

Le Serrec described the monster's mouth as being about 4 feet wide, or 1.2 meters, with a flat lower jaw. Its skin was dark brown or black and appeared to have a rough texture like a shark's skin, and it had lighter brown rings in a sparse pattern down its back. It had pale eyes with slit pupils near the top of its head, small teeth, no fins, no visible gills, and Le Serrec also thought he saw a wound on its tail.

The main photo shows something with what looks like a bulbous head and a long body tapering to a point, although it's just a dark shape in the water. Two other photos were taken at closer range and show the head, including the small pale eyes.

Several famous cryptozoologists investigated, including Bernard Heuvelmans and Ivan Sanderson. They were skeptical but open-minded, which is the best way to approach photographs of gigantic tadpole-shaped sea monsters.

Sanderson suggested that if it wasn't a hoax or something like a big piece of plastic mistaken for a living creature, it might be a giant swamp eel. Swamp eels are freshwater fish that look like eels and live in tropical and subtropical areas. The swamp eel especially likes marshes and ponds, and can breathe air and even wriggle on land to move from one pond to another. Its mouth and throat contain a lot of blood vessels that absorb oxygen and act as a primitive lung. Most swamp eels start life as females and after a few years they change into males, although some start out as males and stay that way.

The swamp eel has smooth skin, almost no fins, and is dark brown. But the largest species, the marbled swamp eel, only grows to about 5 feet long at the most, or 1.5 meters, and it only lives in freshwater. Whatever the Hook Island sea monster was, it wasn't any kind of swamp eel.

As cryptozoologists investigated, though, some iffy things came up. In the late 1950s, Le Serrec had told people he had a project that would bring him a lot of money, something to do with a sea serpent. Five years later, surprise! Here are some sea monster pictures, with Le Serrec shopping them around to magazines to find who would pay him the most for them. It also turned out that Le Serrec was wanted by Interpol, apparently for taking money from people to fund his sea monster expedition in the 1950s.

The photos themselves show something suspicious. Usually the photo you see online is a colorized, sharpened picture of the original photo, the one taken from a distance showing the whole monster. It's pretty spectacular. But the original photos were black and white and not super high quality. In the closer-up photos, the edges of the monster's head look like they're covered in sand—not in the way a real animal's body looks when it's partially buried in sand, but a waviness that looks as though someone had covered the edges of a plastic or cloth bag with sand to weight it down.

From the original photograph

WATER OWL

I f you do a search for mythical animals that turned out to be real, the water owl is on just about every single list. The water owl, also called the Xiphias, is supposedly a huge sea monster with the body of a fish and the head of an owl, with big round eyes. It was supposed to ram ships with its sword-like beak or slice through them with its huge dorsal fin.

According to those solved-mystery-animals lists, supposedly the monster was really a Cuvier's beaked whale. But in T.H. White's 1960 translation of the *Book of Beasts*, a 12th century bestiary, Xiphias is clearly identified as a swordfish. Elsewhere it's also called gladius, "so-called because he has a sharp pointed beak, which he sticks into ships and sinks them." Not coincidentally, the swordfish's scientific name is *Xiphias gladius*, which basically means "sword sword."

Directly under the Gladius entry is that of the serra, which "is called this because he has a serrated cockscomb, and swimming under the vessels he saws them up." I don't know what the serra is supposed to be and neither does T.H. White. It's possible it was a muddled account of the sawfish.

There is no entry for sea owl, water owl, or anything similar in any bestiary I could get my hands on. It's possible that the confusion of the Xiphias of medieval bestiaries and Cuvier's beaked whale comes from the whale's scientific name, *Ziphius cavirostris*. Xiphos means sword in Greek and the whale does have an elongated beak, although nothing like a swordfish's, and not even very long compared to other beaked whales. Another common name for it is the goose-beaked whale, which is a lot more accurate.

Its face and its beak look nothing like an owl's, nor does it have a very big dorsal fin. It grows up to 23 feet long, or 7 meters, and can be gray, brown, or even a reddish color. It's a deep diver and habitually feeds on squid and deep-sea fish. In fact, it holds the record for the longest and deepest recorded dive for any mammal—over two hours underwater and over 9,800 feet deep, or nearly 3,000 meters. That's almost two miles. Its flippers fold back into depressions in its sides to reduce drag as it swims. Like other beaked whales it has no teeth except for two tusks in males that stick up from the tip of its lower jaw. Males use these tusks when fighting and many whales have long scars on their sides as a result.

The swordfish also has no teeth, but it does have a hugely elongated bill that it uses not to spear fish, but to slash at them. It's a fast, scary-looking fish that can grow up to 15 feet long, or 4.6 meters, and it does have a pronounced upright dorsal fin. While its bill isn't exactly owl-like, since owls all have short bills, it does have huge round eyes.

In other words, the water owl isn't Cuvier's beaked whale. It's probably the swordfish.

Like Cuvier's beaked whale, the swordfish spends a lot of time in deep water. Its deepest recorded dive was over 9,400 feet, or 2,865 meters. It eats fish, squid, and crustaceans, swallowing the smaller prey whole and slashing the larger prey up first.

One interesting note about the swordfish is its eyes. Like marlin, tuna, and some species of shark, the swordfish has a special organ that keeps its eyes and brain warmer than the surrounding water. This improves its vision but it's also really unusual in fish, which are almost exclusively cold-blooded. Sometimes a solved mystery is more interesting than an unsolved one.

STELLER'S SEA-APE

Georg Wilhelm Steller was a German botanist and zoologist who lived in the 18th century. In 1740 he was part of an expedition to the Bering Sea between Siberia's Kamchatka Peninsula and Alaska. On the way back from an unscheduled trip to Alaska after one of the ships got lost, they were shipwrecked on what is now called Bering Island, where half the crew died of scurvy. The other half managed to build a boat from the wreckage of their ship and sailed it back to Kamchatka.

During the several years of this expedition, Steller took careful notes on the animals and plants he encountered, including some animals we still can't identify. One of those is called Steller's sea-ape.

Steller only saw the sea-ape once off the Shumagin Islands in Alaska on August 10, 1741, but he did watch it for more than two hours. He described it as a little over 5 feet long, or 1.5 meters.

> The head was like a dog's head, the ears pointed and erect, and on the upper and lower lips on both sides whiskers hung down which made him look almost like a Chinaman. The eyes were large. The body was longish, round, and fat, but gradually becoming thinner toward the tail; the skin was covered thickly with hair, gray on the back, reddish white on the belly, but in the water it appeared

entirely red and cow-colored. The tail, which was equipped with fins, was divided into two parts, the upper fin being two times as long as the lower one, just like on the sharks.

However, I was not a little surprised that I could perceive neither forefeet as in marine amphibians nor fins in their place.[1]

This description sounds like a seal of some kind, but all seals have fore-limbs. One suggestion is that it was a young Northern fur seal, and that either Steller missed seeing its forelimbs or it was an individual born without them. I'm not sure why the suggestion is that it was a *young* seal, though. Baby Northern fur seals are black at birth and lighten to brown as they grow, with older males having some gray patches. All appear black in the water, not reddish. Adult females only grow to about 4.5 feet long, or 1.4 meters, and males about 7 feet long, or 2.1 meters, so at 5 feet long Steller's animal was already a fully grown female or a nearly full-grown male. Young and female Northern fur seals don't have the long whiskers Steller describes, although males do—but only when full grown.

While it's possible Steller's sea-ape was a small male Northern fur seal with no front flippers or flippers that Steller inexplicably didn't see, there is one other issue. Steller would have known perfectly well what a Northern fur seal looked like. They're threatened now due to overhunting and habitat loss, but in the mid-18th century they were plentiful throughout the Bering Sea.

Either Steller saw a Northern fur seal that was so malformed that he didn't recognize it, or he described a different animal. Or, as deep-sea ecologist Andrew Thaler suggests, it was a hoax.

Here's the situation: Steller and the Danish captain of the ship *St. Peter*, Vitus Bering, absolutely did not get along. The expedition was primarily for charting and exploring the region, not describing new animals, and Bering considered Steller nothing more than the ship's physician and a nuisance. When the ship got lost and ended up off the coast of Alaska, Steller had to beg Bering to let him explore this new land. He only got ten hours to do so.

When the crew was stricken with scurvy, Bering initially refused to allow Steller to treat the crew. This was before people understood what vitamins were and while many cures for scurvy were available, no one knew why they worked. When people don't know how something works, sometimes they're suspicious of it.

Steller wrote that Bering was sarcastic, refused his medical advice, and treated him with contempt in front of the crew. Steller's sighting of the sea-ape was only about six weeks after his ten-hour shore leave, so it's reasonable to assume that Steller was still feeling resentful.

Three months later the ship wrecked and the crew was marooned for eight months. Steller spent the time turning his notes into a book. Bering spent the time dying of scurvy. Steller didn't include the sea-ape in his book.

Thaler points out that Steller didn't just name his mystery animal a sea-ape, he named it the Danish sea-ape, *Simia marina Danica*. Bering was Danish, the only Dane on the ship.

The situation is made even more complicated by the name. In his journal, Steller writes, "As for its body shape, for which there is no drawing, it corresponds in all respects to the picture that Gesner received from one of his correspondents and in his *Historia animalium* calls *Simia marina Danica*. At least our sea animal can by all rights be given this name because of both its resemblance to Gesner's *Simia* and its strange habits, quick movements, and playfulness."[2]

The problem with this statement is that the drawing Steller mentions doesn't look anything like a seal. It also doesn't look anything like Steller's description of the sea-ape. It's probable that Steller hadn't seen the picture in a long time and was going by memory, and that he primarily just remembered the name. But the name isn't correct either. Gesner, in his 16th century bestiary, published the drawing and named it *Simia marina*, which means "sea ape." The only Danish thing about it is its artist, who was Danish.

The Simia marina *drawing from Gesner's bestiary*

Steller had a subtle wit and was probably poking fun at Bering in the name, but I'm hesitant to say he made the whole sighting up. The details not only point to a real animal, they aren't malicious. He described the sea-ape playing with some kelp, swimming back and forth under the ship, things like that.

I think Steller sighted a real animal and took notes, probably because he was so bored he would have taken detailed notes on *anything*. Maybe he knew he was watching a Northern fur seal and amused himself by comparing it to Bering. Maybe he didn't know what the animal was but some aspects of it reminded him of Bering. Maybe he left it out of his book because he knew it was a caricature. Maybe he left it out of his book because he didn't have enough information to include it. Maybe he meant to add it later, after he hopefully sighted more of the animals. I don't know.

What I do know, though, is that someone else saw a Steller's sea-ape in June of 1965.

A British man named Miles Smeeton, which was apparently his real name, was sailing his yacht near the Aleutian Islands when he, his wife, his daughter, and a friend all saw a strange animal they couldn't identify. It was around 5 feet long, or 1.5 meters, with reddish-yellow fur, a doglike head, and long whiskers like a shih-tzu. It dived underwater when the ship got near. No one saw any limbs and they were all convinced it wasn't a seal or a sea otter.

So who knows? Maybe there's a limbless seal or some other mammal swimming around in the frigid waters of the Bering Sea, just waiting to be discovered.

~

Steller's Sea-Raven

A BIRD NAMED after Steller has never been definitively identified, Steller's sea-raven. During the winter he spent shipwrecked on Bering Island, Steller wrote in his journal about a bird he called a white sea-raven. He didn't say much about it, just that it was new to him and that it only landed on cliffs along the island's coast so he couldn't get a close look at it. No one knows what bird he was talking about.

A lot of people have made suggestions, of course. One researcher thinks it might be a type of cormorant, since the word for cormorant in German means sea-raven, and in fact that's what the word cormorant means in the original Latin too—*corvus marinus*. Cormorants are black birds with usually small white markings, so the cormorant Steller saw might be a white or mostly white species that is now extinct. Then again, the bird might have been something else entirely.

Since we don't have more to go on than this brief description of a white sea-raven that likes ocean-facing cliffs, it's hard to know what to look for.

∽

Steller's Sea-Cow

STELLER'S SEA-COW WAS A SIRENIAN, related to dugongs and manatees. Sirenians evolved around 50 million years ago and share a common ancestor with elephants. Their front flippers have toenails that look like elephant toenails, which is neat. They're fully aquatic and, like both whales and seals, they're mammals that breathe air. They live in shallow water and graze on aquatic plants. Occasionally they do eat a jellyfish, but who hasn't accidentally eaten a jellyfish? The sirenians living today grow to around 13 feet long at most, or 4 meters, but Steller's sea-cow was more than twice that length, up to 30 feet long, or 9 meters.

Steller's sea-cow was a type of dugong and had a whale-like notched tail instead of a rounded tail like a manatee's. Instead of teeth, it had chewing plates made of keratin that it used to chomp lots and lots of kelp and other plants. Its hide was thick with a thick layer of blubber underneath to keep it warm in the cold water. It had a long upper lip covered in bristles that helped it grab plants, and its forelegs were flipper-like and small.

Steller discovered the sea-cow while he was shipwrecked on Bering Island in 1741. It lived there and around some of the other islands in the Bering Sea, although fossil and sub-fossil remains indicate it used to be much more widespread.

Unfortunately, once it was discovered by Europeans, it was hunted to extinction within thirty years, killed for its oil-rich blubber and for food. But Steller's sea-cows have occasionally been spotted after that, although no one has provided actual proof. Many of the sightings may be of hornless female narwhals, which live in the area and are about the same color and shape as the Steller's sea-cow when seen from the surface.

In 1962, some whalers spotted six animals in shallow water off the coast of Kamchatka, and whalers can probably be relied upon to recognize a whale when they see it. These animals looked like dugongs. In 1976, a sea-cow carcass reportedly washed up on shore not that far from where the whalers' sighting had taken place. Some workers at a nearby salmon factory went out to look at it and described it as more like a dugong than a whale, but no one thought to keep the body.

After that there were a couple of expeditions to look for surviving Steller's sea-cows. While none were found, the coast of Kamchatka and its numerous islands are rugged and hard to explore. The sea-cow might still be hanging on.

NINGEN

The seas around Antarctica are cold and stormy. To humans it seems unhospitable, a deadly ocean surrounding an icy landmass. But the Antarctic Ocean is home to many animals, from orcas and penguins to blue whales and colossal squid, not to mention the migratory birds, cold-adapted fish, and many small animals that live in the depths. New animals are constantly being discovered, but it's also not very well explored.

Stories from Japanese whalers who visit the area supposedly tell of a strange creature called the ningen, which is occasionally seen in the freezing ocean. It's usually white and can be the size of a big person or the size of a baleen whale. It's long and relatively slender, and while details vary, it's generally said to have a human-like face, or at least large eyes and a slit-like mouth. It also has arms instead of flippers and either a whale-like tail or human-like legs.

These stories don't come from long ago, though. The first mention of the ningen anywhere appeared in 2002 in a Japanese forum thread about giant fish. The topic drew a lot of interest initially but that died down within a few months, until 2007 when the ningen was the subject of both a manga and a magazine article.

The ningen didn't start appearing in English language sites until 2010.

While it's never been as well-known as many so-called cryptids, it has been the subject of short stories and books, creepy art, a J-pop song, and lots of speculation.

The question, of course, is whether the ningen is a real animal or a hoax. The initial post was made by an anonymous woman who claimed to be repeating something an unnamed whaler friend told her he'd experienced, *and* her friend also said that the Japanese government was baffled, *and* that the government was engaged in a cover-up so no one else would learn about the mystery animal. This has all the hallmarks of a modern urban legend. I don't think the ningen is a real animal.

Just for fun, though, if it was a real animal, what might it be? The beluga whale is the first thing I thought of, since it's white, grows around 18 feet long, or 5.5 meters, and has a small rounded head with features that look sort of human-like. But the beluga whale only lives in the Arctic, not the Antarctic. That's the opposite side of the world.

Of the whales that do live around the Antarctic for at least part of the year, none are white all over and most are dark gray or black. Very rarely, though, a whale is born with albinism, which means its skin lacks pigment. As a result, it looks white or very pale gray. An albino humpback whale called Migaloo has been spotted off the coast of Australia repeatedly since 1991, for instance.

An albinistic bowhead or right whale living in the Antarctic might be seen occasionally by whalers who don't realize they're all seeing the same individual. Both the bowhead and right whales have deep, rounded rostrums that could potentially look like a human face—slightly, if you were looking at it through fog or darkness and were already aware of the story of the ningen.

Then again, if the ningen is a real animal, it might be a whale that's completely unknown to science. There are still a lot of beaked whales we know almost nothing about, and new species of beaked whale are occasionally discovered. The ningen might not even be a whale at all but something else entirely.

Still, while it's a fun story, it's probably not real. You can't believe everything you read on the internet.

PART FIVE
MYSTERY
CARCASSES

If you can't get enough of sea monsters, here's a section about (mostly) sea monsters that are doubly mysterious and triply disgusting!

CANVEY ISLAND MONSTER

C anvey is a 7 square mile, or 18.5 square kilometer, island off the southern coast of England not far from London. It's barely above sea level and on January 31, 1953, a tidal surge overtopped the sea wall in the night and drowned 58 people. Its marshes are home to lots of plants and animals, including some insects that at one point were thought extinct. It was also a fashionable vacation area in Victorian times and can claim lots of ghosts, such as one story told by night fishermen who sometimes see a Viking standing on the mudflats staring out to sea. He supposedly drowned while waiting for his ship to return.

Canvey Island's big claim to fame these days is something that happened late in the same year of the big flood, 1953.

This is the story as reported pretty much everywhere. Sometime in November of 1953, a dead animal washed ashore. We don't know exactly what day it was or who found it. It was lying in shallow water and its finders pulled it farther ashore and covered it with seaweed, presumably so nothing would bother it and it wouldn't wash back out with the tide. They went for the police, but the police had no idea what they were looking at. They called "the government," who sent two zoologists to identify the body. The zoologists didn't know what it was either. They had the body incinerated and left without making an official report.

The body measured 2.5 feet long, or about 76 centimeters. It's described as a marine animal with thick brownish-red skin, protruding eyes in a pulpy head, sharp teeth, and gills. It also had hind legs but no front legs. Remarkably, its feet each had five toes that together were shaped roughly like a horseshoe. The zoologists reportedly said it looked as though it would be able to walk upright on its legs.

Then, in summer of 1954, another one washed ashore. This one was bigger, almost 4 feet long, or 120 centimeters. It weighed about 25 pounds, or 11.3 kilograms. A short article appeared on August 13, 1954 in either the Canvey *Chronicle* or the Canvey *News*. The headline reads "Fish with feet found on beach."

> A fish with feet was found on the beach at Canvey on Tuesday by the
> Rev. Joseph D. Overs. He described the fish as being over four feet
> long with staring eyes and a large mouth. Underneath, on its stom-
> ach, it had two feet, each with five toes. It was dead and had appar-
> ently been damaged by being washed against the rocks. A peculiar
> fish was found in almost the same place last year and identified as a
> pocket or 'fiddler fish.'

Under that is the subheading "SEAL TOO" and the sentence "For the first time within living memory a seal was seen in Benfleet Creek, near the bridge, on Tuesday."

There's a lot to unpack in this story, not to mention that pocket fish and fiddler fish seem to be local names and don't match the common names for any fish. There's a fiddler ray, sometimes called a banjo ray, which I'm delighted to learn is a type of guitarfish. Guitarfish are only slightly guitar shaped. They mostly look like little sharks if you smooshed the shark's head flat. The fiddler ray has a rounder flattened head than a guitarfish. It lives around Australia and likes shallow, sandy bays, where it mostly eats shell-fish and crabs. It's harmless and edible. But it's not reddish-brown, it doesn't have sharp teeth, and it certainly doesn't have anything that could be called legs by any stretch of the imagination.

It also seems odd that the newspaper article doesn't mention the two zoologists supposedly sent by "the government" who couldn't identify the 1953 monster. For that matter, it doesn't say that the 1954 fish was the same type of thing found in 1953. It just says "a peculiar fish was found in almost the same place last year." Not the same kind of fish. The same *place*.

As it happens, you don't have to look too hard to find out how this got so scrambled. In 1959—only about five years after the weird thing washed ashore on Canvey Island—writer and radio personality Frank Edwards published a book called *Stranger Than Science*. It's since been reprinted many times and I have clear memories of reading it as a kid, although I don't remember anything about the Canvey Island monster. It was a popular book and full of less than stellar research.

Edwards' book is the main source used for subsequent accounts of the Canvey Island monster. It's Edwards who claims there were two such monsters, Edwards who describes the feet as having toes arranged in a U shape, Edwards who introduces us to the mysterious government-sent zoologists who tell everyone the monster is a bipedal marine animal but it's okay, it's harmless, hey, let's just burn this body and *tell no one*. It appears that Edwards made a lot of this up.

The 1954 newspaper story was picked up by the Associated Press, but the full text of the AP article is even shorter than the original, although slightly more sensational:

> A grotesque sea creature four feet long and with two five-toed feet
> was found on the beach here Tuesday by Reverend Joseph D. Overs.
> He described the thing, which was dead, as 'a sort of fish with
> staring eyes and a large mouth underneath. It has two perfect feet,
> each with five pink toes.'

The original 1954 article says that Reverend Joseph D. Overs found the body. According to the CanveyIsland.org page, while Overs was a reverend, he wasn't the local vicar or anything like that. Apparently he was a reverend of the Old Roman Catholic Church of Great Britain, with a handful of parishioners who met for services at his lodging house. He was better known as the island's photographer and was popular and well-liked. He took the photo of the fish himself, although he may not actually have been the one to find it. The webpage suggests that the reporter included Overs' title of reverend to give the article more zing and that Overs didn't usually use his title.

The CanveyIsland.org site is for residents, with a chatty tone, and many of the comments are from people who knew Overs. One 2011 comment about the mystery fish monster, left by someone named Colin Day, reads:

> I was THERE. I was a young lad of nine at the time. I noticed a group of peers in a crowd on the beach. Kids were prodding it with their spades. I ACTUALLY TOUCHED IT! I thought it was a person at first as I could only see part of it through the crowd. Its flesh was NOT fish-like scales. It was a pinkish color and looked like wobbly human flesh with cellulite, orange peel texture. I remember shouting to the other kids 'It's a mermaid' over and over.

While the fish itself is long gone—no one's sure what happened to it, but a deep hole in the sand was probably involved, because I bet it stank—we do have a single black and white photograph. What does it show?

It's a wide-bodied fish with a huge gaping mouth, fins or projections of some kind to either side, and a long, tapering tail. Since it's a face-on photo, it's hard to get a good idea of where the fins are situated. They seem to be near the massive head but might be farther back. The fish appears pale, at least in

Not exactly fashion — and certainly not a gift, except, perhaps to an angler. Pictured on Canvey Beach. What is it?

comparison to the dark ground, and we have the eyewitness description of at least one little boy that it was pink, although Edwards claims it was reddish-brown.

Locals are convinced it was an anglerfish. Ichthyologists have suggested an anglerfish species known as a monkfish or a related species called a frogfish.

The monkfish is broad and flattish, with a tapering tail, a big wide mouth with sharp teeth, and two roughly triangular fins jutting out from its sides. It lives in the ocean around England as well as in the Mediterranean and Black seas. It hunts among seaweed near the ocean floor, sometimes using its muscular fins to walk itself along instead of swimming. Its skin doesn't have scales but it is bumpy. Like other anglerfish, it has a lure on its head, a highly modified dorsal fin spine, that it can move around to attract small fish and other prey. When something touches the lure, *yomp*! The monkfish gulps it down. It has an expandable stomach and can swallow prey as big as it is, and it can get big—almost 7 feet long for a big female, or over 2 meters.

The frogfish prefers tropical and subtropical oceans. It's smaller than the monkfish, barely more than a foot long or around 35 centimeters, and it's rounded rather than flattened. Some species of frogfish have elaborate filaments all over their bodies that help them blend in with seaweed and other plants. The frogfish frankly doesn't look much like the fish in the picture, and is too small to fit the description, but it does have one thing in the plus column that the monkfish doesn't. Many species are orange, yellow, or pink in color. The monkfish is dark.

There are more than 200 species of anglerfish known and many are seldom seen because they live so far down on the bottom of the ocean. It's definitely possible there are unknown species and that one washed ashore on Canvey Island in 1954. It's too bad no one kept the fish, but at least we have a photo.

Or do we? We don't actually know that that photo accompanied the 1954 article. The Canvey Island library has archives of one of the two newspapers from that era—but the 1954 papers are missing. Writer Garth Haslam, who has researched the Canvey Island monster extensively, points out that the original description of the fish doesn't mention its tail, which is quite long and would have been notable. He suggests the picture may actually accompany a different article entirely, one from 1967.

The Londonderry *Sentinel* of August 12, 1954, ran the same article that appeared in other newspapers, but with one very important addition at the end:

> The fish, which is also known as angler, sea devil, frog or toad fish, and fishing frog, is a British fish, and the name Angler is said to have been derived from its preying on small fish, which it attracts by moving worm-like filaments attached to the head and mouth.

Even if the newspaper picture didn't come from the 1954 article, it seems clear from this article that we're talking about anglerfish anyway. Even the 1953 fish's identification as a fiddler fish isn't too surprising, since the fiddler ray does superficially resemble an anglerfish in that it has a large head but a much more slender body that tapers in a long tail. The anglerfish's fins are strong and thick, and if the body was damaged as Overs reported, the ends of the fins may have been frayed to resemble toes.

In *Stranger Than Science*, Frank Edwards describes the fish as having five toes arranged in a U shape. Where on earth did that come from? Well, for some reason, Edwards was convinced that the Canvey Island monster was the same thing that left hoofmarks in the snow all over Devonshire in February of 1855. No one else has made that connection and I have no idea why Edwards decided to link them. Devon and Canvey are over 200 miles apart, or about 360 kilometers. If Edwards wanted to use the Canvey Island monster to solve the mystery of the devil's footprints, he had to make people believe not only that the fish was bipedal but that it had feet whose prints would resemble hooves.

Anglerfish are fascinating animals but I promise you, they aren't running around on land leaving little hoofprints (but you can read about what was in the "Demons and Specters" section of this book).

GLOBSTERS

I f you live near the seashore, or really if you've spent any time at all on the beach, you'll know that stuff washes ashore all the time. You know, normal stuff like jellyfish that can sting you even though they're dead, pieces of debris that look an awful lot like they're from shipwrecks, and the occasional *solitary shoe with a skeleton foot inside*. But sometimes things wash ashore that are definitely weird. Things like globsters.

A globster is the term for a decayed animal carcass that can't be identified without special study. Globsters often look like big hairy blobs and are usually white or pale gray or pink in color. Some don't have bones but some do. Some still have flippers or other features, although they're usually so decayed that it's hard to tell what they really are. And they're often really big.

In 1924, off the coast of South Africa, witnesses saw a couple of orcas apparently fighting a huge white monster covered with long hair—far bigger than a polar bear. It had an appendage on the front that looked like a short elephant trunk. Witnesses said the animal slapped at the orcas with its tail and sometimes reared up out of the water. This went on for three hours.

The battle was evidently too much for the monster and its corpse washed ashore the next day. It measured 47 feet long in all, or 14.3 meters, and the body was 5 feet high at its thickest, or 1.5 meters. Its tail was 10 feet long, or over 3 meters, and its trunk was 5 feet long and over a foot thick, or about 35 centimeters. It had no legs or flippers. But the oddest thing was that it didn't seem to have a head either, and there was no blood on the fur or signs of fresh wounds on the carcass.

The carcass was so heavy that a team of 32 oxen couldn't move it. The reason someone tried to move it was because it stank, and the longer it lay on the beach the worse it smelled.

Despite its extraordinary appearance, no scientists came to investigate. After ten days, the tide carried it back out to sea and no one saw it again. Zoologist Karl Shuker has dubbed it Trunko and has written about it in several of his books.

When a shark decomposes, it can take on a hairy appearance due to exposed connective tissue fibers. But Trunko was fighting two orcas only hours before it washed ashore.

Or was it?

No one saw the fight from up close and orcas are well known to play with their food. There's a very good chance that Trunko was already long dead and that the orcas came across it and batted it around in a monstrous game of water volleyball. That would also explain why there was no blood associated with the corpse.

In that case, was Trunko a dead shark? At nearly 50 feet long, it would have had to be the biggest shark alive...and as it happens, there is a shark that can reach that length. It's called the whale shark, which tops out at around 46 feet, or 14 meters, although we do have unverified reports of individuals nearly 60 feet long, or 18 meters—or even longer.

The whale shark is a filter feeder and its mouth is enormous, some 5 feet wide, or 1.5 meters, but the interior of its throat is barely big enough to swallow a little fish. Its teeth are tiny and useless. Instead, it has sieve-like pads that it uses to filter tiny plants and animals from the water. The whale shark either gulps in water or swims forward with its mouth open, and water flows over the filter pads before flowing out through the gills. Tiny animals are directed toward the throat so the shark can swallow them.

The whale shark is gray with light yellow or white spots and stripes and has three ridges along each side. Its sandpaper-like skin is up to 4 inches thick, or 10 centimeters. It has thick, rounded fins, especially its dorsal fin, and small eyes that point slightly downward. It usually stays near the surface but it can dive deeply too, and it's a fast swimmer despite its size. It isn't dangerous to humans at all, but humans are dangerous to it since poachers kill it for its fins, skin, and oil even though it's a protected species.

The whale shark usually lives in warm water, especially in the tropics, but occasionally one is spotted in cooler areas. It's well known off the coast of South Africa. If the Trunko globster was a dead whale shark, the "trunk" was probably the tapered end of the tail, with the flukes torn or rotted off. Most likely the jaws had rotted off as well, leaving no sign that the animal had a head or even which end the head should be on.

But sharks aren't the only big animals in the ocean, and once the skin and blubber of a dead whale have broken down sufficiently, the collagen fibers within them can look like fur. Collagen is a connective tissue and it's incredibly tough. It can take years to decay. Tendons, ligaments, and cartilage are mostly collagen, as are bones and blubber.

While we don't know what Trunko really was, many other globsters that have washed ashore in modern times have been DNA tested and found to be whales. For instance, in 1990 the Hebrides blob washed ashore in Scotland. It was furry and 12 feet long, or 3.7 meters, with a small head at one end and finlike shapes along its back. Despite its weird appearance, genetic analysis revealed it was a sperm whale, or at least part of one. Another sperm whale revealed by DNA testing was the Chilean blob, which washed ashore in Los Muermos, Chile in 2003. It was 39 feet long, or 12 meters.

The St. Augustine monster was found by two boys bicycling on Anastasia Island off the coast of Florida in November 1896. It was partially buried in sand but after the boys reported their finding, people who came to examine it eventually dug the sand away from the carcass. It was 21 feet long, or almost 6.5 meters, 7 feet wide, or just over 2 meters, and at its tallest point was 6 feet tall, or 1.8 meters. Basically, though, it was just a huge pale pink lump with stumpy protrusions along the sides.

A local doctor, DeWitt Webb, was one of the first people to examine the carcass. He thought it might be the rotten remains of a gigantic octopus and described the flesh as being rubbery and very difficult to cut. Another witness said that pieces of what he took to be parts of the tentacles were also strewn along the beach, separated from the carcass itself.

Dr. Webb sent photographs and notes to a cephalopod expert at Yale,

Addison Verrill. Verrill at first thought it might be a squid, but later changed his mind and decided it must be an octopus of enormous proportions—with arms up to 100 feet in length, or over 30 meters.

DeWitt Webb standing on the St. Augustine monster

In January a storm washed the carcass out to sea, but the next tide pushed it back to shore 2 miles away, or 3.2 kilometers. Webb sent samples to Verrill, who examined them and decided it was more likely the remains of a sperm whale than a cephalopod.

The tissue samples of the St Augustine monster still exist and they've been studied by a number of different people with conflicting results. In 1971, a cell biologist from the University of Florida reported that it might be from an octopus. Cryptozoologist Roy Mackal, who was also a biochemist, examined the samples in 1986 and also thought the animal was probably an octopus. A more sophisticated 1995 analysis published in the *Biological Bulletin* reported that the samples were collagen from a warm-blooded vertebrate—in other words, probably a whale. The same biologist who led the 1995 analysis, Sidney Pierce, followed up in 2004 with DNA and electron microscope analyses of *all* the globster samples he could find. Almost all of them turned out to be remains of whale carcasses, of various different species.

Sometimes a globster is pretty obviously a whale, but one with a bizarre and unsettling appearance. The Glacier Island globster of 1930, for instance, was found floating in Eagle Bay in Alaska, surrounded by icebergs from the nearby Columbia Glacier. The head and tail were skeletal but the rest of the body still had flesh on it, although it appeared to be covered with white fur.

Its head was flattish and triangular and the tail was long. The men who found the carcass thought it had been frozen in the glacier's ice.

They hacked the remaining flesh off to use as fishing bait but they saved the skeleton. A small expedition of foresters came to examine it and reported that it measured 24 feet and one inch, or over 7.3 meters. They identified it as a minke whale. The skeleton was eventually mounted and put on display in a traveling show, advertised as a prehistoric monster found frozen in a glacier. In 1931 the skeleton was donated to the National Museum of Natural History in Washington DC, where it remains in storage. Modern examinations confirm that it's a minke whale.

In May 2007 a huge, peculiar-looking dead animal washed ashore in Guinea in Africa. It looked like a badly decomposed alligator of enormous size, with black plates on its back that almost looked burnt. It had a long tail and legs but it also had fur. Its mouth was huge but there were no teeth visible.

It wasn't fur, of course, but collagen fibers, but what's up with the burnt-looking plates on its back? That's actually not rare in decomposing whales. The carcass is lying on its back so the plates are actually on the belly. Its ventral pleats are even visible in photos, which are what allows a whale to expand its mouth as it engulfs water before sieving it out through its baleen.

So yes, it was a dead baleen whale, and we even know what kind. The legs aren't legs but flippers, and details of their shape and size immediately let whale experts identify this as a humpback whale.

Another strange sea creature, referred to as the Ataka carcass, washed ashore in Egypt in January 1950 after a colossal storm that didn't let up for 72 hours. When the storm finally abated, a huge dead animal was on the beach. It was the size of a whale and looked like one except that it had a pair of tusks that jutted out from its mouth. Witnesses said it had no eyes but they did note the presence of baleen.

The baleen identified it as a whale and the tusks were just bones exposed by the stormy water. They're called mandible extensions and the whale itself was identified as a Bryde's whale. In baleen whales the lower jaw is made of two separate bones called mandibles, mandible extensions, or just lower jaws. They're only loosely attached and often separate after death, especially after being tossed around in a storm.

The longest Bryde's whale ever measured was just under 51 feet, or 15.5 meters. It's related to blue whales and humpbacks and mostly eats small fish, cephalopods, and other small animals. It's a swift, slender whale, the only baleen whale that lives year-round in warm water, so it doesn't need blubber to keep it warm.

Many globsters have stumps that look like the remains of flippers, legs, or tentacles. In 1988 a treasure hunter found a globster now called the Bermuda blob. It was about 8 feet long, or almost 2.5 meters, pale and hairy with what seemed to be five legs. The discoverer took samples of the massively tough hide, which were examined by Sidney Pierce in his team's 1995 study of globster remains. This was one of the few that turned out to be from a shark instead of a whale, although we don't know what species.

But sharks don't have five legs. The leg-like protrusions are probably flesh and blubber stiffened inside with a bone or part of a bone, such as a rib. As the carcass is washed around by the ocean, the flesh tears in between the bones, making them look like stumps of appendages.

There's a good reason why so many globsters turn out to be sperm whale carcasses. A sperm whale's massive forehead is filled with waxy spermaceti oil. The upper portion of the head contains up to 500 gallons of oil in a cavity surrounded by tough collagen walls. Researchers hypothesize that this oil is used both for buoyancy and to increase the whale's echolocation abilities. The lower portion of the forehead contains cartilage compartments filled with more oil, which may act as a shock absorber since males in particular ram each other when they fight. Much of the head of a sperm whale, which can be one-third of the whale's length, is basically a big mass of cartilage and connective tissue. After a whale dies, this buoyant section of the body can separate from the much heavier skeleton and float away on its own.

Globsters aren't a modern phenomenon. We have written accounts of what were probably globsters dating back to the 16th century, and older oral traditions from folklore around the world. The main problem with globsters is that they're not usually studied. They smell bad, they look gross, and they may not stay on the beach for long before the tide washes them back out to sea.

After Hurricane Fran passed through North Carolina in 1996, a group of young men found a globster washed up on a beach on Cape Hatteras. They took pictures and estimated its length as 20 feet long, or 6 meters, 6 feet wide, or 1.8 meters, and 4 feet high at its thickest, or 1.2 meters. From the pictures it's pretty disgusting, like a lump of meat with intestines or tentacles hanging from it.

The men weren't supposed to be on the beach, which was part of the Cape Hatteras National Park and closed due to hurricane damage. They didn't mention their find to anyone until the following year, when one of the men learned about the St Augustine monster in his college biology class. By then, of course, the Cape Hatteras globster was long gone. While it might have been a rotting blob of whale blubber or a piece of dead shark, we don't know for sure. So if you happen to find a globster on a beach, make sure to tell a biologist or park ranger so they can examine it before it's lost to science forever.

MINNESOTA ICEMAN

Thi s book isn't about Bigfoot—it's right there in the title—but we *are* going to talk about one Bigfoot-type creature. It's called the Minnesota iceman.

It happened in 1968, although the iceman had been around for a while as part of a traveling exhibition. In 1968 a zoologist named Terry Cullen saw it at a livestock fair in Chicago. Curious and excited, Cullen talked to the owner of the exhibition, named Frank Hansen. Hansen said he wasn't actually the iceman's owner but that he was taking care of it for the owner.

Cullen contacted two biologists who were interested in cryptozoology and urged them to look at the iceman, Ivan Sanderson and Bernard Heuvelmans. Both wrote several influential books and were right in the thick of cryptozoological things back in the day. Ivan Sanderson in particular managed to have all sorts of sightings of mystery animals with absolutely no proof except for his own say-so.

But the iceman was clearly a real thing, because there it was. It wasn't a living animal but a dead one preserved in a block of ice. It was apelike but much larger than most apes with human-like proportions, and would have stood about 6 feet tall when alive, or 1.8 meters. The iceman's face was somewhat flat in shape with pronounced brow ridges and a short snub nose. It had dark fur and enormous hands and feet, and while the ice obscured a lot of details, both Heuvelmans and Sanderson got a good close look at it in Hansen's home that December. The exhibition was on its winter break and Hansen kept it at his house in a refrigerated case. The body itself showed signs of injury, including a possibly broken arm and an eyeball dislodged from its socket.

Both Sanderson and Heuvelmans took photos and made careful drawings of the iceman. Heuvelmans published a description of it in a Belgian scientific journal in 1969. He first suggested it was a new species of human that he named *Homo pongoides*, but later he changed his mind and decided it was probably more closely related to Neanderthals.

Sanderson, meanwhile, was active in drumming up media interest. He was also sure it was a new species, but instead of *Homo pongoides*, he chose to call it Bozo.

Sanderson published an article about the iceman in a 1969 issue of *Argosy* science fiction magazine, where he was an editor. In that article he suggests the iceman was the missing link between humans and apes. He also contacted John Napier, a primatologist who worked for the Smithsonian Institution and had a longstanding interest in the Yeti and other Bigfoot-type creatures, and suggested he investigate. Napier and the Smithsonian's director, S. Dillon Ripley, also a Yeti enthusiast, were both interested. But when they asked to see the iceman, suddenly Hansen said nope, the iceman's actual owner didn't want anyone else looking at their... sideshow exhibit.

Later Hansen said Napier and Ripley could see it after all. But when they did, it was clear to them that it wasn't a real animal. Napier said it looked like a latex model. Hansen promptly said that he'd hidden the original, worried that he would get in trouble if it turned out to be a type of human. Sanderson and Heuvelmans did say that the photos Napier took looked different from their own photos and what they remembered. But Hansen wouldn't let anyone see the original iceman, so the scientists' interest waned and that was pretty much it.

So where did the iceman come from and where is it now? Surely by now some museum or university has bought the actual specimen and thawed it out to examine, right? Where's the genetic profile? Why isn't this thing in textbooks?

Let's start with where it came from. Hansen said the body had been discovered in Siberia by seal hunters, floating in a block of ice just off the coast. But he also said it had been found not by Russian seal hunters but by Japanese whalers. Okay, well, maybe he couldn't remember exactly who found it. But wait. He also said it had been shot by a hunter in Minnesota. And he also said it was killed in Vietnam and smuggled into the United States during the war in a body bag. And he also said someone discovered it in Hong Kong in a deep freeze after it was confiscated by authorities from the seal hunters or possibly whalers.

In other words, Hansen was clearly just making all this up. But maybe he didn't know where it came from. After all, he always said he wasn't the iceman's owner, although he wouldn't say who the owner was, just that they were an "eccentric California millionaire." He also claimed the specimen's hairs and blood had been tested and came back with lots of anomalies, but he refused to let anyone have current samples and he conveniently didn't have copies of the test reports.

In 2013 someone listed the iceman in an eBay auction. It was bought by the Museum of the Weird in Austin, Texas, where it's still on display. It looks like the photos and illustrations made by Heuvelmans and Sanderson, but it's not a real animal. It's a latex model, just like Napier suspected.

Back in the early 1970s, Napier had done some digging and discovered that in 1967, Hansen had commissioned a latex model of an ape-man from a company that made movie props.

It's quite possible that there was more than one model of the iceman made, since obviously that would increase its earning power in sideshow exhibits. That would explain why the iceman Heuvelmans and Sanderson saw looked different from the iceman Napier and Ripley saw. Or the differences could have come about when the model was thawed for easier storage, then frozen again in ice when it was time to go back on tour.

Of course, some people refuse to believe that the iceman was a hoax. They believe Hansen really did switch out the actual real iceman with a model. They base this on the report by Heuvelmans and Sanderson that when they examined the iceman, there was a place where the ice had melted, exposing the skin, and they said they smelled putrefaction.

There are several possible explanations for this, however. Just as people see what they expect to see, people smell what they expect to smell. I'm sure that model didn't smell great, either, especially after being exhibited all over the country all summer. My freezer smells kind of funny even though there's literally nothing in it but ice. Any questionable whiffiness might have been interpreted as that of putrefaction. Or, of course, Hansen might have thrown a piece of rotten meat in its case to make it smell more realistic.

Whatever the reason for the smell, that's a really slim reason to ignore all the evidence that points to the Minnesota iceman being a latex model. I mean, we actually have the model to look at, and we know Hansen commissioned a model right before the Minnesota iceman conveniently went on exhibition.

Everything points to a hoax, but you can't blame Hansen. He didn't approach Sanderson and Heuvelmans, they approached him. Hansen was a showman, so he probably hoped that if he convinced the pair the iceman was real, it would lead to a lot more publicity and therefore more money for him. That was his job, after all.

MONTAUK MONSTER

On July 12 or 13, 2008, depending on which source you consult, three friends visited Ditch Plains Beach, near the little town of Montauk, New York, in eastern North America. It was a hot day and the beach was crowded, and when the three noticed people gathered around something, they went to look too. There they saw a weird dead animal that had obviously washed ashore. One of the three took a picture of it, which appeared in the local papers and on the local TV news. From there it went viral and was dubbed the Montauk monster.

The monster was about the size of a cat but with shorter legs, a chunkier body, and a relatively short tail. It didn't have much hair but it did have sharp teeth, and the front part of its skull was exposed so that it almost looked like it had a beak. Its front paws were elongated with long fingers, almost like little hands.

People all over the world tried to guess the monster's identity, everything from a sea turtle without a shell to a diseased dog or just a hoax. Some people thought it was a mutant animal that had been created in a lab on one of the nearby islands. They said it must have escaped and died trying to swim to the mainland.

While no one knows what happened to the animal's body, scientists have studied the photo and determined that it was probably a dead raccoon that had been washed into the ocean. The waves had tumbled the body around through the sand long enough to rub off most of its remaining fur and some of its facial features, and then it washed ashore during the next high tide. It was also somewhat bloated due to gases building up inside during decomposition. It's the animal's teeth and paws that made the identification possible, since both match a raccoon's exactly.

~

Island View Carcass

IN MAY OF 2019, a couple out for a walk came across a rotting carcass washed up on Island View Beach in British Columbia, Canada. It was over 5 feet long, or 1.5 meters, and had tough dark brown skin, a reptilian snout, and thorn-like hooks along its long tail. It didn't look like any known animal, although it was so decomposed that details were hard to make out.

Some people thought this might be a small Cadborosaurus, the sea monster that's supposed to live in nearby Cadboro Bay. Others thought it was a kind of shark due to its shark-like skin.

Once the story appeared in the local news, someone contacted the media with photos taken of the carcass a few weeks earlier. Not only was it less rotten then, there were two other carcasses nearby that looked very similar.

Brian Timmer, a biologist at the University of Victoria, analyzed the photos and determined that the animals were longnose skates. They didn't look like skates because all three were missing their wings. Skates have broad winglike pectoral fins that they use to glide through the water. Researchers speculate that someone caught the skates and cut the wings off to eat, then threw the animals back into the water where they died and washed ashore.

The longnose skate is a type of ray that lives along the northeastern Pacific coast of North America. It eats worms, crustaceans, and other small animals it finds on the sea floor. It can grow up to 6 feet long, or 1.8 meters, with a long tail and a pointed nose that gives it its name. Rays are related to sharks, which explains why some people thought it was a shark. It's not really big enough or scary enough to be called a sea monster, though, unless you're a worm.

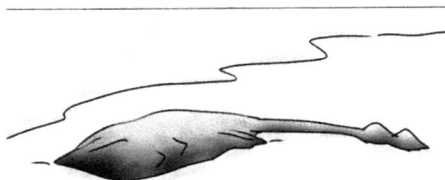

From the photograph taken in May 2019

TECOLUTLA MONSTER

In the town of Tecolutla in Veracruz, Mexico in 1969, some locals walking along the beach at night saw a monster in the water. It was 72 feet long, or 22 meters, with a beak or fang or bone jutting from its head, huge eye sockets, and hair-like fibers covering its body. Some witnesses said it was plated with armor too. It was floating offshore and later the people who found it claimed it was still alive when they first saw it. Since the hairy fibers are a sign of a whale or shark that's been dead and decomposing in water for considerable time, they probably mistook the motion of the carcass in the waves for a living animal swimming.

The locals who found the carcass thought its bones were made of ivory and would be valuable. They kept their find a secret for a week and managed to haul it onshore. It took them 14 hours.

They tried to cut it apart on the beach but only managed to remove the enormous head. By that time the rest of the body was starting to get buried in sand.

At that point the locals, frustrated, decided they needed heavy machinery to move the thing. They told the mayor of Tecolutla that they'd discovered a crashed plane, probably expecting the city to send out a crane big enough to move a plane and therefore big enough to move their monster. But, of course, when the volunteer rescue party showed up to the supposed plane crash, all they found was a really stinky 72-foot-long corpse.

The mayor decided that a stinky 72-foot-long corpse was exactly what tourists wanted to see. Instead of hauling it out to sea or burying it, he moved it in front of the town's lighthouse so people could take pictures of it.

He was right, too. A college student who traveled to the town to film the event said there were a hundred times more tourists in the area than usual, all to look at the monster.

Biologists eventually identified it as the remains of a sei whale. The horns or beak were probably jaw bones. And in case you're wondering, ivory is a component of teeth, not bones.

PART SIX
OUT OF PLACE
ANIMALS

A kangaroo in Australia is fine. A kangaroo in North America is out of place. In this section we'll meet lots of out-of-place animals, from phantom kangaroos in various parts of the world to big cats in England to the bonnacon of South America.

PHANTOM KANGAROOS

O ver five days in January 1934, an animal described as kangaroo-like was repeatedly seen outside the town of South Pittsburg, Tennessee, near Chattanooga. It reportedly killed several dogs, including large police dogs, along with geese, ducks, and other livestock. One witness said it "looked like a giant kangaroo running and leaping across the field." It was large, brown in color, and extremely fast, but that's the only description given. A search party followed strange tracks up Signal Mountain, but lost the tracks near a cave.

Newspapers from other parts of the country thought the idea of a kangaroo leaping around killing dogs in rural East Tennessee was *hilarious*. They trotted out all the stereotypes of mountain men and moonshine, until the local papers got defensive and jabbed back with editorials.

When a pair of hunters killed an animal described as a lynx or wildcat on Signal Mountain, the local papers decided that was the culprit. There's even a photo of the poor dead animal in the Chattanooga *Daily Times*, and it's clearly a bobcat. Bobcats are about twice the size of domestic cats. They certainly look nothing like kangaroos. The bobcat weighed 40 pounds, or 18 kilograms, and measured 50 inches long, or 127 centimeters. That's big for a bobcat, but witnesses who saw the kangaroo said the wildcat looked nothing like it and was definitely nowhere near the size of the animal they saw. They estimated that the kangaroo weighed around 150 pounds, or 68 kilograms.

The local police chief launched a "mad dog drive," where any dogs suspected of being exposed to rabies were killed. By February of 1934 sightings had ceased, but whether the kangaroo was a rabid dog or a bobcat, or something else, we don't know. The mystery was never solved.

Kangaroos and their smaller relations, wallabies, started to appear in American zoos in the early 20th century. Zoos in those days were grim places, with animals kept in small cages and little to no information about them available. The kangaroo was considered an exotic animal but many people had no idea what they ate or how they acted in the wild, or even what a kangaroo looked like. As a result, when an unknown animal was spotted near South Pittsburg, Tennessee, the word kangaroo got attached to it as a way to indicate it was strange and seemingly didn't belong there.

Sometimes kangaroos and wallabies really do appear where they don't belong. There are even entire populations of wallabies living in various places outside of their native Australia, and we don't actually know how they got there in most cases. All we can do is guess they were escaped pets or escaped from zoos.

Wallabies look like miniature kangaroos, maybe the size of an average dog. They eat all kinds of plants, can bound quickly and jump fences and other obstacles, and are largely nocturnal although they also come out in daylight. They fill a similar ecological niche as deer. They breed well in captivity and are cute, so they're often kept in zoos. They're marsupials native to Australia and New Guinea, and introduced to New Zealand...but they're also found in parts of England, Scotland, Ireland, on the Isle of Man, in one tiny area of France, and a few other places.

We do know where some of the wallabies come from. Lady Fiona of Arran was a bit of an eccentric who kept what would have been considered exotic pets back in the 1920s and 30s: llamas, alpacas, pot-bellied pigs, and wallabies. At some point in the 1940s she turned the wallabies loose on an island in Loch Lomond, where they've lived ever since. The problem, of course, is that they shouldn't really be there. They're cute, sure, and tourists do come to see them, but some people think their presence threatens certain native bird species and they ought to be killed off, or at least captured and taken to zoos. Other people point out that the native birds on the island don't seem to be bothered by the wallabies, which after all have been there for the better part of a century now. About sixty wallabies live on the island at any given time.

There are also wallabies in the Peak District in Derbyshire, England. We know where these wallabies came from too. A man named Henry Courtney Brocklehurst, which honestly does not even sound like a real person's name, once kept a private zoo on the Roaches, a rocky ridge that's part of a national park in England. At some point in the 1930s five wallabies in his zoo either escaped or were quietly turned loose. Reports vary. They did well in the wild and multiplied, until at their most numerous there were around fifty. For a while people thought the wallabies had all died off after some especially cold winters, but in 2009 a hiker got pictures of one.

But wallabies also live in other parts of the British Isles, and we don't know where most of them came from. At least some are probably escaped pets or escaped from zoos—wallabies are apparently pretty good at escaping—while others may be individuals from the Peak District population that have gone wandering. All the wallabies in Britain are a subspecies of red-necked wallaby called Bennett's wallaby, which is from Tasmania and better adapted to the British climate. They're sometimes killed by cars or dogs.

So let's get back to more modern sightings of phantom kangaroos. There are a lot of reports from the United States and a few from Canada, including a 1949 sighting in Ohio, a 1958 sighting in Nebraska, the "Big Bunny" kangaroo sightings in Minnesota that persisted for a decade between 1957 and 1967, and many more.

On October 18, 1974 someone called the Chicago police to report a kangaroo on their porch. Two officers responded and found the kangaroo in a nearby alleyway. They reported it was 5 feet tall, or 1.5 meters, which points to its being an actual kangaroo and not a wallaby. Chicago cops probably are not trained to deal with kangaroos, but you'd think they would have the sense to call animal control or a zoo. Instead, they decided to treat the kangaroo like a human criminal.

That's right. They tried to handcuff a kangaroo.

The kangaroo kicked and punched the officers, who called for backup and probably still haven't lived that down, but the kangaroo jumped a fence and disappeared into the night.

The next day a paperboy near Oak Park reported hearing a car's brakes squeal, and when he turned to look, he saw a kangaroo staring at him. It hopped away. Over the next few weeks sightings poured in from all over Chicago, from other Illinois cities, and even from Indiana. The kangaroo was never caught and by July 1975, sightings had tapered off and finally stopped.

In April 1978, a school bus driver in Waukesha, Wisconsin saw two kangaroos hop across the road. She wasn't the only one to see them, either. The road was busy and drivers had to slam on their brakes to avoid the kangaroos. One driver actually hit one, but it just jumped back up and hopped away. More people spotted the kangaroos in the weeks that followed. By the end of the month, a kangaroo hunt was organized in an attempt to capture it. They failed and it seems mostly to have been a joke anyway. An anonymous photographer sent a Polaroid of a kangaroo to the local paper, supposedly taken in the area. It turned out to be a photo of a stuffed wallaby someone had dragged out into a field. But sightings of the kangaroo persisted for months.

Sightings of kangaroos and wallabies don't just happen in North America and the British Isles. Between 2003 and 2010, people in the Mayama mountain district of Osaki, Japan reported seeing a large brown animal with long ears, variously described as 3 to 5 feet tall, or 92 to 152 centimeters. Most sightings of the animal took place on roadsides when people were on the way to and from work.

No one has pictures of the kangaroo seen in Japan, though. No one has photos of the kangaroos reported in Chicago or any of the other sightings from 1934 on up through the new millennium. All those kangaroos were seen but never caught or photographed. According to a century of reports, the phantom kangaroos are just that, phantoms.

Until recently.

On January 3, 2005, a woman in Iowa County, Wisconsin called the sheriff's department to say she'd seen a kangaroo hopping around on her horse farm. When the sheriff arrived, he was shocked to actually find a kangaroo, specifically a male red kangaroo. He knew better than to try to handcuff it and instead got help to lure the kangaroo into a barn. It was captured safely and taken to the Henry Vilas Zoo, where it stayed until it died five years later. No one knows where it came from or why it was loose, but it was apparently fairly tame.

In 2013, some hunters got video of a kangaroo in Oklahoma. A pet kangaroo named Lucy Sparkles had escaped in November of 2012 not too far away, but it turned out that the kangaroo caught on video wasn't Lucy. It also wasn't an escapee from a local exotic animal farm. It was a kangaroo from a man who kept a few as pets, which means there are way more kangaroos in Oklahoma than I ever imagined.

Similarly, a man driving to work in July 2013 in North Salem, Connecticut got a few seconds of video of a kangaroo or wallaby bounding down the road and up someone's driveway. In February of 2017, police officers on patrol spotted a wallaby in Somers, New York, which is believed to be a pet wallaby that escaped three years before in North Salem.

People really are seeing wallabies and kangaroos, not phantoms. Is it possible there are populations of the animals living in remote areas of North America, and that occasionally one wanders closer to a town or city and is seen? It's possible but not especially likely. Wallabies would be easy prey for wolves and coyotes, bears and cougars, and domestic dogs. We'd also probably have a lot of roadkill wallabies. Kangaroos would be more able to hold their own against predators, but are larger and therefore easier to spot. The climate in much of the United States also isn't ideal for kangaroos or wallabies.

I'm pretty sure that the wallabies and kangaroos seen in North America, and probably most other places, are escaped pets. Wallabies in particular seem to be escape artists. So if you're tempted to get an exotic pet and are considering a wallaby, trust me: you'd be a lot happier with a dog.

BONNACON

In the early 16th century, the Italian scholar Antonio Pigafetta sailed around the world with Portuguese explorer Ferdinand Magellan. Technically he only sailed part of the way around the world with Magellan, who died in the Philippines. Pigafetta wrote a detailed account of the voyage after he returned to Spain in 1522.

Pigafetta reported that the natives of Patagonia told him about devils with two horns and long hair that breathed fire and also farted fire. These interesting fire details aren't reported by anyone else, so it's possible that Pigafetta added them to make the story better. He also would have been familiar with the bonnacon, an animal found in bestiaries at the time and written about by Pliny the Elder.

The bonnacon was described as a bull with a long mane like a horse and horns that curled backwards. Because its horns couldn't be used for defense, it was supposed to run away from danger and fart so prodigiously that the fumes would set fire to everything nearby and poop would be scattered across three acres. Medieval bestiaries played this for laughs but people also believed it. It's possible that Pigafetta thought the Patagons were describing the bonnacon. It's also likely, incidentally, that the bonnacon was a type of buffalo or bison, many of which have small curved horns. Many hoofed animals will void their bowels when stampeding away from predators, so this could be the start of the story.

While the farts of flame seem to be Pigafetta's invention, many Patagonian tribes do have stories of horned animals and spirits that seem remarkably bovine. In the late 19th and early 20th century, a man named Lucas Bridges collected many traditional stories of the people in Tierra del Fuego, which is at the very tip of Patagonia and which is remote even now, and was certainly remote a century ago. He reported that the Selknam people told stories about Hachai, a horned man with white fur and red stripes who acted as a fierce and powerful protective spirit along with his two sisters. Bridges witnessed a pantomime of Hachai that was a remarkable imitation of cow-like behavior, but the man performing it had never even seen a cow. Because here's the thing: there is no known bovid native to South America.

If the Selknam had never seen the cattle introduced by the Spanish, and there are no native bovids in Patagonia, how did they imitate cattle so perfectly? In 1833, in southern Chile, a man of the Chono tribe visited a ship and while there, he saw two powder horns. He put them to his head and bellowed like a bull. Moreover, while in much of South America the local native languages borrow the word for cattle from Spanish, native Patagonian languages have their own words for cattle.

There are two theories. The first has to do with a shipwreck. In 1540 a ship belonging to the Bishop of Plasencia's fleet sank in the Strait of Magellan. It carried livestock and we know that some sheep survived. A 1557 expedition reported sheep in the area and in 1741 some natives brought three freshly killed sheep to the leader of another expedition. It's possible that some cattle survived long enough to make an impression on the local population, and many stories of horned water monsters have been collected in Patagonia. But if we take Pigafetta's report of the fire-farting horned spirit as inspired by cattle sightings, the shipwreck happened a few decades too late.

The other theory suggests that there was once a bovid that lived in Patagonia. There are a few small hints that this may have been the case. A

1586 Spanish document refers to a buffalo-like animal with "horns with their tips curved backwards which this witness guesses must be buffalo." In 1598 explorer Oliver van Noort reported animals like stags and buffalo at Puerto Deseado. The stags were actually guanaco, which are related to llamas, but we don't know what his buffalo might have been.

Bovids originated in Eurasia and entered North America relatively late, and as far as researchers can tell none ever made it farther south than Mexico until domestic cattle were brought to South America by the Spanish. By the mid-1500s cattle had been introduced into the Pampas, a vast prairie north of Patagonia, and feral herds may have made their way to Patagonia by the end of the century.

The Spanish cattle were tough and adaptable, and a small population still lives wild in the Andes. They have adapted to life in forests and to bitterly cold weather, including growing long fur in winter. No one reports that they fart fire.

ALIEN BIG CATS

There are so many reports of big cats in places where big cats should not be, especially black panthers, that cryptozoologists have a term for them: ABCs, or Alien Big Cats. Big cat reports from Great Britain are especially common.

Many reports of black panthers turn out to be large domestic cats seen at a distance, where tricks of perspective and poor light contribute to the cat looking much bigger than it really is. Some big cat reports turn out to be real animals that escaped from captivity, but other reports are not so easy to explain away.

The British Isles only fully separated from mainland Europe about 8,000 years ago. Before that it was connected to what is now Denmark and the Netherlands via a large area of marshland for several thousand years. We know one European big cat crossed into Britain during that time, the Eurasian lynx, which only went extinct in Britain about 1,500 years ago.

The Eurasian lynx still lives in parts of Europe and Asia. It's larger than the related North American bobcat and the Canadian lynx, but still nowhere near as big as the cave lion that lived in Britain and other areas until it went extinct around 11,000 years ago. The Eurasian lynx stands about 28 inches tall at the shoulder, or 70 centimeters, and is heavily built with thick spotted fur and a short bobtail. The tip of its tail is black although the rest of the animal is mostly buff to orangey-brown with darker brown spots, and it has long black tufts of fur on the tips of its ears.

The first modern report of an Alien Big Cat in Britain comes from about 1770. The writer William Cobbett described seeing a gray cat the size of a Spaniel dog when he was a boy. It climbed into a hollow elm tree near Waverley Abbey, a ruin in Surrey. Later in his life, while traveling in Canada, he saw a wild gray cat, possibly a lynx, and thought it looked like the cat he saw in England.

Lynxes are occasionally killed or captured in Britain. In 1903 a lynx was shot in Devon after it killed two dogs. The animal was taxidermied and given to a local museum, which labeled it as a Eurasian lynx—but a study of the animal published in 2013 proved that it was a Canadian lynx that had been kept in captivity for at least part of its life.

In 1927 newspapers reported a lynx caught in Scotland but didn't give much information about the animal. It's only called large, fierce, and yellow, and the shepherds who'd seen it before it was trapped said it looked like a leopard without spots. That's strange if this really was a lynx, because while some individual lynxes may have spots that don't show up well against the background coat, all lynxes have spots. They also have bobtails and the distinctive black ear tufts, but those details aren't reported.

It's possible the poor cat wasn't a lynx at all but a puma. The puma, also called a cougar or mountain lion, is native to the Americas. It's bigger than a lynx, standing almost 3 feet tall at the shoulder, or 90 centimeters, and is usually tawny in color with lighter belly. You'd think that the London Zoo expert who examined the animal would know the difference between a long-tailed puma and a short-tailed lynx, though, so who knows what it really was?

In 1980 an honest-to-goodness puma was captured in Scotland. The puma spent the remainder of her life at the Highland Wildlife Park zoo, where they named her Felicity. After she died she was taxidermied and is now in the Inverness Museum. Felicity was probably an escaped or released pet, since the zoo director reported that she liked being tickled. Wild animals don't typically like to be tickled.

Many other large cats of various species have been killed or captured in

Britain. In most, if not all, cases the animals were probably exotic pets that were either released when they became too difficult to manage, or escaped and weren't reported because the owners didn't have a license to own the animal in the first place. But there are many other sightings of more mysterious large cats.

The Surrey Puma is one of the earliest cases. The first sightings were made in the 1930s around the Surrey/Hampshire border, but since this is the same area where William Cobbett saw his Spaniel-sized gray cat, many people think the 1770 account is the earliest known. In 1955 a woman walking her dog saw a puma-like cat slinking away from a dead calf, which it had evidently been feeding on. Many reports followed throughout the 1960s. Naturalists who examined paw prints discovered in the area identified the prints as made by dogs. Occasionally someone would snap a photo but they were always too blurry to be conclusive—until 2014.

One quiet Friday morning a man named Allan Tinkler, from Molesey in north Surrey, was eating breakfast with his children when he saw a strange cat in his garden. It stayed in the yard for over half an hour and was definitely not a domestic cat. The pictures he took clearly show a small wildcat called a serval.

The serval is a long-legged, large-eared African cat with spots over most of its body and some stripes along the neck and shoulders. It turned out that this particular serval was an escaped pet, and its owner retrieved it safely and took it home. They also had a license to keep it.

Obviously that serval wasn't the cause of sightings going back decades or even centuries, but with no clear photos or captured animals, we can't guess what the Surrey Puma might really be.

The Beast of Bodmin Moor is another case that received a lot of attention starting in the late 1970s. Bodmin Moor is in Cornwall, England, and over the years people have reported seeing long-tailed big cats in the area, brown or black in color and about the size of a German shepherd. One persistent rumor is that in 1978, when a circus owner named Mary Chipperfield had to close her zoo in Plymouth, she released three pumas onto the moor rather than give them to Dartmoor Zoo. Supposedly the Beast of Bodmin Moor is a descendant of these pumas.

In 1994, after something killed a number of sheep on the moor, the local

MP called for an official investigation into the beast. A biologist and a zoologist headed the investigation. They looked at the sheep supposedly killed by a big cat, at tracks, photographs, and video footage, and conducted a search of the moor. Their conclusion, after six months, was that there was no evidence for any big cats of any type living on the moor. The tracks were made by dogs, the sheep carcasses did not show signs of being killed or eaten by cats, and the photos and videos showed domestic cats seen in places where scale was hard to judge.

Locals weren't at all appeased. Many people were convinced big cats lived on the moor. A few days after the official report, a boy found the skull of a big cat near the River Fowey. But the Natural History Museum in London, which examined the skull, determined that it was from a leopard that had been prepared for taxidermy after death and was probably taken from a leopard-skin rug.

Sightings of big cats continue in Britain, although without any definitive proof that they aren't just misidentified domestic cats or other known animals. If there are big cats in Britain, they're most likely escaped individuals and not a breeding population.

∾

Wildcats of Madagascar

MADAGASCAR IS a large island off the coast of Africa, home to lemurs and other animals found nowhere else in the world. It doesn't have any native felids, although people who live on Madagascar do have pet cats. But a scientist named Michelle Sauther, who researches lemurs, kept seeing cats in the forest. They were all tabbies and the locals called them wildcats, but Dr. Sauther wanted to know more about them.

She and her team set up traps for the forest cats. When they trapped a cat, they took photographs, hair and blood samples, and even dental impressions. Then they released the cats back into the wild. Genetic profiles developed from the samples helped solve the mystery of what these cats are. They're feral cats descended from ship cats that traveled from areas around the Arabian Sea, hundreds of years ago and possibly as much as a thousand years ago. Enough cats jumped ship on Madagascar to develop into a breeding colony that is still around today.

Dr. Sauther is studying the effects of the feral cats on local animals, because cats can cause a lot of damage as introduced predators. Cats are efficient hunters of small animals, especially rodents like mice, but also birds, reptiles, amphibians, insects, and basically any animal they can catch, especially the fuzzy kind stuffed with catnip.

MISPLACED CAMELS

T he camel is an even-toed ungulate but it doesn't have hooves. Instead it has padded feet with two hoof-like toenails. There are three species alive today but the one we're going to talk about here is the dromedary, which has only one hump and is native to the Middle East and the Sahara in northern Africa. The dromedary no longer lives in the wild except as a feral animal and hasn't for a few thousand years. It was domesticated at least 4,000 years ago.

In that case, though, why are there feral camels where no camels should logically be: namely, Australia and possibly North America?

The dromedary is a beast of burden, which is a fancy way of saying people use it to carry heavy loads and to ride. As a result, it's been introduced to other parts of the world, especially parts of Asia and southern Europe. Those introductions mostly happened a long time ago, as much as 1,000 years ago or more in some areas and some 600 years in others. It's much more recently that people decided Australia and North America needed camels. After all, parts of the interior of Australia and parts of the southwestern United States and Mexico are deserts, and camels are adapted to live in deserts.

The first camel in Australia was supposedly imported in 1840, a single camel named Harry who settled in to life down under successfully, although he was known as an ornery cuss even by camel standards. Harry was used as a pack animal until he was sold to a man named John Horrocks, a British explorer.

In 1846 Horrocks was shooting birds on a lakeshore in South Australia near the end of an expedition. It's not clear exactly what happened, but Harry was kneeling next to Horrocks and moved or started to get up just as Horrocks was reloading his shotgun. The pack Harry was carrying jostled against the gun and it went off, right in Horrocks's face. Horrocks died of his wounds a few weeks later, but his dying wish was for Harry to be shot dead. So Harry, the first camel in Australia, was killed in September 1846 and Horrocks will forever be known as the man who was shot by a camel.

It wasn't until 1860 that camels started being imported to Australia in great numbers. So the timeline here is: no camels ever in Australia until 1840, then one camel from 1840 to 1846, then no camels again, and finally a bunch of camels starting in 1860.

So why is there a report from 1846 of a random camel wandering around in Australia in the early 1800s?

The report comes from a book called *A Visit to the Antipodes: With Some Reminiscences of a Sojourn in Australia*, which was written by a man called E. Lloyd. We don't know anything about E. Lloyd except that he wrote this book about his travels. We don't even know his first name, although I bet it was Edward.

Lloyd says that a camel had been imported by someone in the North Country but that it had escaped and was living in the bush. Occasionally it would appear seemingly out of nowhere, frighten people who had never seen a camel before, which was most people in that area at that time, and wander off again. Eventually a donkey who had also escaped from a farm made friends with the camel and they were often seen together.

If this is true and not just something Lloyd made up, it means at least one camel was brought to Australia before 1840. But we don't know who or whether there were more. We also don't know that camel's name, but it might have been Edward too, who knows?

Camels were popular for transport and carrying goods across Australia for decades, with many cameleers brought in from India, Turkey, Egypt, and other countries to handle and care for the animals. In the 1920s, trucks started becoming more common in Australia and by the 1930s the camels were on the way out. Many camel owners just released their animals into the wild once they were no longer useful. The camels were like, yeah fine, whatever, we don't even need humans, and formed feral herds. Some of them were recaptured by Aboriginal people, who didn't have cars and trucks at the time and were pleased to get some free camels so they could visit relatives in distant villages more easily.

The problem, of course, is that camels don't belong in Australia. They're an invasive species and they're very successful. Not only are there no predators of adult camels in Australia, there are no camel diseases since imported camels were carefully chosen to be disease-free. By 2013 there were an estimated 600,000 feral camels in Australia. They aren't as damaging to the environment as some invasive species, but they do take resources away from native animals and damage fencing and water pumps used for livestock.

Between 2009 and 2013, Australia killed around half of the feral camels in the country. The program was expensive and funding was ended in 2013, but after a series of droughts and fires, a new camel cull started in 2020. The goal isn't to kill all the camels, which would be nearly impossible since there may be as many as a million camels by now, but to reduce their numbers so camels aren't destroying waterholes that native animals depend on.

On the other side of the world, an ancestor of modern camels once lived in North America. It appears in the fossil record around 45 million years ago and was extremely successful. Its descendants lived throughout North America, including in the Arctic. It also migrated into South America and evolved into today's llamas and their relations, and into Eurasia by the Bering land bridge, where it evolved into Bactrian and dromedary camels. North America not only used to have camels, it used to have the first camel.

But the last camels of North America went extinct around 11,000 years ago when so many of the other ice age megafauna also went extinct...until 1857, when the United States decided that what the American west needed was camels. They imported 75 dromedaries from the Middle East to carry supplies between military outposts.

The camels were scattered throughout the west, from Texas to California, and were also used to carry loads during expeditions along the Mexican border. Apparently they did a good job and were well suited to the environment, but most soldiers didn't know how to handle camels and didn't like them as a result. They tried to treat them like mules, which didn't work at all. By 1863 some of the camels had been turned loose, some had escaped, and some had been sold to circuses, zoos, and a few farmers.

Meanwhile, though, other people were also importing camels to use as beasts of burden. Some miners used camels to carry their packs, since they were stronger than mules and could withstand the harsh conditions of the desert better. But by the time trains and trucks became commonplace, the camels had mostly already been killed for meat, turned loose to fend for themselves, or just died off.

There never were as many camels imported to North America as to Australia and they were scattered widely throughout the west, so they probably didn't form feral herds. But dromedaries can live to be about 40 years old. There were credible sightings of camels in various places until around 1890.

That brings us to the legend of the Red Ghost. The Red Ghost was a hideous devil that rode on the back of a monstrous animal. The Ghost had been sighted throughout what was then the Arizona territory, and stories of it spread throughout the mining camps and ranches in the area. The animal it rode on was huge and dangerous, rumored to have trampled people to death and killed horses and bears. No one knew whether the rider was a devilish person or a person-like devil.

Then two things happened that solved that mystery only to create a new one. A group of miners saw the Red Ghost along a river and, terrified, they shot at it. They missed the animal but hit the rider, who fell to the ground and lay still. When the miners approached they found a partially mummified skeleton that had fallen off the animal's back.

Later, a rancher shot and killed an animal he recognized as a camel. When he examined the body he found strips of rawhide that had been used to lash the rider to the camel so securely that it hadn't fallen off until the miners shot the body off. No one knew who the person might have been or why he had been tied to a camel.

That story is creepy, but it's also probably not true. A rancher did kill a feral camel in Arizona in the mid-1880s when he caught it eating his garden, though, which is not nearly as creepy.

PART SEVEN
CREEPY-CRAWLIES

Not every animal is a cuddly, pettable mammal. In this section we learn about the Mongolian death worm, the Tatzelwurm of Europe, and lots more, including a mystery fossil from 300 million years ago.

TATZELWURM

The tatzelwurm story comes from many parts of the Alps, and while the story is different in different places, the animal is basically a snake with a catlike head. Belief in the tatzelwurm's existence has been documented as far back as 1680 although the folklore is probably much older. Some stories say it has poisonous breath, can jump long distances to attack people who approach it, and that it steals milk from cattle, all common details in European dragon stories.

The tatzelwurm is generally described as a snake or lizard, gray or whitish in color with large bright eyes, a rounded head often described as catlike, and a stubby tail. Sometimes it's described as having no legs at all, only front legs, or the normal complement of four legs. Occasional stories say it has multiple legs beyond four.

There have been a lot of sightings over the years in different areas. The earliest well-documented sighting is from around 1711, when a father and son named Tinner spotted one on a mountain in Switzerland. They described it as coiled up on the ground, limbless, but with a catlike head. It was dark gray and about 7 feet long, or over 2 meters. In 1779 a man named Hans Fuchs saw two tatzelwurms, ran home in a panic and told his family, and promptly dropped dead of a heart attack.

More recent sightings mostly come from the early 20th century. In 1921, a poacher and his friend hunting in the Alps in Austria saw a tatzelwurm on a rock, watching them. It was gray, 2 or 3 feet long, or 61 to 91 centimeters, with two legs and a catlike head. Its body was described as thick as a human arm and its head was fist-sized. The poacher shot at it but it jumped at the men and they ran. In 1922, two sisters playing in the woods in St. Pankraz, Austria saw a strange creature crawling among some rocks. It was gray, a bit over a foot long, or 30 centimeters, and looked like a giant worm with a pair of paws behind its head. In 1929, an Austrian schoolteacher hiking in the Alps spotted a snake-like animal that he estimated was 18 inches long at most, or 45 centimeters, pale and smooth, with a flattened head with large eyes, and two small front legs he described as atrophied-looking.

Occasionally someone reports finding a dead tatzelwurm, but as is common in cryptozoological reports, the body vanishes before it can be examined by experts. In 1828 a peasant in Switzerland found a tatzelwurm corpse in a dried-up marsh. He collected it but before he could send it to be examined, crows ate it. He did eventually send the skeleton to Heidelberg, but it either never arrived or was lost when it got there.

Over the centuries, stuffed specimens, dead animals, reptile skins, and photographs have been offered as proof of real tatzelwurms, but every case has proved to be either a hoax or misidentification of a known animal. The Alps are a huge mountain range some 750 miles long, or 1,200 kilometers, crossing eight countries. They're not as high as the Andes, Himalayas, or Rockies, but like those mountain ranges, they have plenty of unexplored, hard to reach areas. Plants and animals new to science are still occasionally found there, including a new species of viper only discovered in 2016.

At one point the Nature and Forestry departments in Austria insisted that the tatzelwurm was just an otter, but locals scoffed at this. They knew what otters looked like and this was no otter.

The tatzelwurm does seem to like wet areas. In fact, in some parts of the Alps its appearance is said to be an omen of flooding. Some people have suggested it might be a salamander. Amphiumas are snake-like salamanders that live in water and have vestigial limbs, and the siren salamanders are eel-like with four limbs. The greater siren can even grow to almost 3 feet long, or 91 centimeters. But all these salamanders are only found in North America and none of the known sightings of the tatzelwurm really fit the description of an amphibian.

The amphisbaenian is a better fit. They're burrowing reptiles, usually no bigger than about 6 inches long, or 15 centimeters, that eat insects and earthworms. They resemble earthworms although they have scales. Most species have blunt heads and blunt tails so that it's hard to tell which end is which, and since their eyes are tiny and almost invisible, that adds to the confusion. But the tatzelwurm's eyes are frequently referred to as being large and bright.

A skink, on the other hand, might fit the description. Skinks are long, slender lizards with large eyes and somewhat rounded heads. Many species have no legs and most species with legs have very small ones. In 2012, a skink with a pair of tiny front legs and no hind legs was discovered in Thailand, and the common burrowing skink from South Africa only has hind limbs.

Most skinks like to burrow. Some even dig elaborate tunnel networks. A skink will frequently come out to bask in the sun but will flee to its burrow if disturbed. This sounds a lot like the tatzelwurm. Skinks are generally bigger than amphisbaenians too. The Solomon Islands skink is over a foot long, or 30 centimeters, not even counting its tail. So it's not out of the question that the tatzelwurm might be a skink that can grow several feet long, maybe as much as a meter, and either has no hind limbs or quite small ones that are easily overlooked. Skinks with reduced limbs move like snakes, not like worms.

There don't seem to have been very many organized searchers for the tatzelwurm. In 1997, cryptozoologist Ivan Mackerle led an expedition in the Austrian Alps without any luck. He also interviewed locals about the tatzelwurm and discovered that only older residents knew about it. He suggested it may be extinct. Since the expedition was only a week long, I think he might be jumping the gun a little bit.

MONGOLIAN DEATH WORM

The story goes that a huge wormlike creature lives in the western or southern Gobi Desert. Most of the time it stays below ground, but during the rains of June and July it sometimes comes to the surface. It's generally described as looking like a sausage or an intestine, red or reddish in color, as thick as a person's arm, and as long as 3 or 4 feet, or up to about 1.2 meters. Its head and tail look alike, sort of like a giant fat earthworm, although some reports say it has some bristles or spines at one end.

Touching a death worm is supposed to lead to a quick, painful death. Some people say it can even spit venom or emit an electrical shock that kills people or animals at a distance.

Parts of the Gobi are less dry than you'd expect from a desert. The ground can be quite moist under the dry surface and there are actually earthworms that live in some areas. Two new species were only described in 2013. The earthworms don't resemble reports of the Mongolian death worm, but if an earthworm can survive, other soft-bodied creatures can too. That's assuming that the death worm is actually a worm and not a reptile or amphibian of some kind.

The best suggestion for what the death worm might be is an animal called the amphisbaenian. It's sometimes also called the worm lizard, and

while it's not any kind of lizard, it is a reptile. Amphisbaenians live in many parts of the world, including most of South America and parts of North America, parts of Africa, southern Europe, and the Middle East. Since amphisbaenians live almost all of their lives underground, it's very likely that species unknown to science live in other places. Much of the Gobi is remote, sparsely populated by nomadic herders, and not very well explored by scientists.

Amphisbaenians resemble snakes but they also resemble worms. The eyes are tiny and can be hard to spot, so the head and tail look very similar as a result. Many species are pink or reddish in color, although some are blue or other colors, including spotted. Many have scales that grow in a ringed pattern that make them look even more like earthworms. But they're not big animals, generally around 6 inches long at most, or 15 centimeters, and slender, not as big around as someone's arm. They're also completely harmless to humans and large animals.

That doesn't mean there can't be a big amphisbaenian living in the remote parts of the Gobi, rarely seen even by the people who live there. Or, of course, the Mongolian death worm might be a completely different kind of animal, one totally unknown to science—maybe one that's related to the amphisbaenian but radically different in appearance. It might be a mythical monster instead, although there are enough plausible-sounding witness sightings to think there's something in the Gobi that looks like a big fat red horrible worm, even if it's not actually dangerous.

The death worm isn't the only weird wormlike creature reported from Mongolia. The camel's tail worm is supposedly smaller than the death worm and gray in color, known in the southern part of the country. This animal might actually be the same thing as the camel's tail snake, a sand boa that grows up to about 2 feet long, or 61 centimeters.

In 1981 a Mongolian herder reported seeing a dead worm-like animal near a well, which he described as looking like a salami with wings on the rear of its body. A salami, of course, is a type of sausage that's usually red in color. In 1982, the driver of a scientific expedition in the Gobi noticed something moving under the sand. When he pushed the sand away to look, he saw a beige animal with spade-like paws at its rear, although he didn't see the paws move. The animal quickly reburied itself in the sand. It's likely that both animals were the same thing, and that the herder thought the spade-like rear legs were strange wings.

A species of reptile with only one pair of legs isn't actually all that unusual. Many reptiles have reduced limbs or no limbs at all, including legless lizards that resemble snakes. An extinct relative of modern snakes and lizards had tiny front legs but larger hind legs that it probably used to help it swim. One genus of skinks from Australia, usually called sliders, vary considerably even though they're all closely related. Some species have four ordinary lizard legs while others have tiny legs or no legs at all. In general, species of reptile that burrow have longer bodies with reduced or no legs. Some species of legless lizards in the family Pygopodidae have reduced hind legs, including Burton's legless lizard. It can grow more than 3 feet long, or 92 centimeters, and lives in forested areas of Australia and Papua New Guinea. It has small hind legs but no front legs. One family of amphisbaenians found in Mexico has small front legs that are sometimes mistaken for ears, but no rear legs.

There's a good chance that something strange is living in the sands of the Gobi. Let's hope scientists in Mongolia find it soon so we can learn more about it.

FURY WORM

Carolus Linnaeus was a Swedish botanist who lived in the 18th century. Botany is the study of plants. If you've ever tried to figure out what a particular plant is called, you can understand how frustrating it must have been for botanists back then. The same plant can have dozens of common names depending on who you ask.

Before Linnaeus worked out his system of binomial nomenclature, botanists and other scientists tried various different ways of describing plants and animals so that other scientists knew what was being discussed. They gave each plant or animal a name, usually in Latin, that described it as closely as possible. But because the descriptions sometimes had to be really elaborate to indicate differences between closely related species, the names got unwieldy—sometimes nine or ten words long.

Linnaeus sorted this out first by sorting out taxonomy, or how living creatures are related to each other. It seems pretty obvious to us now that a cat and a lion are related in some way, but back in the olden days no one was certain if that was the case and if so, how closely related they were. It's taken hundreds of years of intensive study by thousands upon thousands of scientists and dedicated amateurs to get where we are today, not to mention lots of technological advances. But Linnaeus was the first to really attempt to codify different types of animals and other organisms depending on how closely they appeared to be related, a practice called taxonomy.

Linnaeus's system is beautifully simple. Each organism receives a generic name, which indicates what genus it's in, and a specific name, which indicates the species. This conveys a whole lot of information in just two words. To make things even clearer, a subspecies name can be tagged on the end.

Linnaeus was a young man when he started working out his classification system. He was only 25 when he traveled to Lapland on a scientific expedition to find new plants and describe them for science. This was in 1732 so travel was quite difficult. Linnaeus traveled on horseback and on foot, which as you can imagine took a long time and gave him lots of time to think. Within three years he had worked out the system we still use today.

You know what else Linnaeus invented? The index card. He needed index cards to keep track of all the animals and plants he and other scientists named using his binomial nomenclature system.

Linnaeus named a whole lot of plants and animals himself—something like 10,000 of them during his lifetime. Naturally enough, some mistakes crept in that have since been corrected. One of these mistakes may be an animal called the fury worm.

First, though, let me tell you something that happened to Linnaeus before he'd even come up with his system of nomenclature. This happened in 1728, when he was a broke college student staying with a professor and spending all his free time collecting botanical specimens in the marshes.

One day Linnaeus was searching for plants he didn't already have specimens of when something stung him on the neck. Since he was wading around in a marsh, this was not really that unusual. But before long Linnaeus's neck was painfully swollen and soon one of his arms had swollen up too.

These days we'd recognize this as an allergic reaction, but back in 1728 they didn't know what allergies were. By the time Linnaeus got home, he was in such bad shape that the doctor they called worried he wouldn't survive.

Fortunately for Linnaeus, and for science and humanity in general, he survived and went on to invent his naming system only eight years later. Some thirty years after he almost died, he published the tenth edition of his book, *Systema Naturae*, and included a formal description of the animal that had almost killed him. He named it the fury worm, *Furia infernalis*.

But there was no type specimen of a fury worm for other scientists to study. Linnaeus hadn't seen the one he believed bit him, and the only one anyone had shown him was a tiny worm so dried up and old that he couldn't see any details. But he knew the fury worm existed because of his own brush with death, and anyway everyone knew it was a real animal.

The fury worm was supposed to be tiny and slender, so small that it could be picked up by the wind and blown to other places. If it landed on a person or an animal it would immediately bite them with its sharp mouthparts, breaking the skin, then burrow into the flesh through the wound. It would dig in so quickly and so deeply that it was impossible to find, and even if you did find it, it was impossible to get out because of the backward-pointing bristles on its tail that kept it anchored in place. A person or animal bitten by the worm was likely to die within a day, sometimes within half an hour, unless a poultice of cheese or curds was applied to the bite.

Fortunately for most of the world, this horrible worm only lived in swampy areas in northern Sweden and Finland, Russia, and a few other nearby areas. In one year, 1823, some 5,000 reindeer died from fury worm attacks and the export of reindeer furs was banned so the worm wouldn't spread.

But. Where. Are. The. *Worms*? And why would a parasitic worm kill its host so quickly? A parasite depends on its host staying alive for enough time that the parasite can benefit from whatever it's getting from the host, whether that's nutrients or a protected place to develop into its next life stage.

We have lots of anecdotal evidence of the fury worm's existence, including from such noted a scientist as Linnaeus himself, but no worms. And the symptoms reported from fury worm attacks varied quite a lot from patient to patient.

Doubts about the fury worm's existence were already common in the 19th century, and even back in the late 18th century Linnaeus started to have doubts. As technology and scientific knowledge improved, the fury worm started to look less and less like a real animal and more and more like an explanation for things people had once not understood—like allergies, infection, and bacteria. The death of 5,000 reindeer in 1823 was finally traced to a disease called neurocysticercosis, which is actually caused by a parasite, but not a fury worm. It's caused by tapeworm larvae that only kill its host after the larvae have matured and are ready to infect a new animal, which happens when something eats the meat of the animal that has died.

So was the fury worm ever a real animal? Almost certainly not, and thank goodness for that.

ICE WORM

T he ice worm sounds like something totally made up, but not only is it real, there are at least 77 species that live in northwestern North America, specifically parts of Alaska, Washington state, Oregon, and British Columbia.

It looks like a tiny earthworm, usually black or dark brown. Some species live in snow and among the gravel in streambeds, while some actually live in glaciers. Unlike the earthworm and most other worms, it requires a temperature of around 32 degrees Fahrenheit, or zero Celsius, to survive. You know, freezing. But the ice worm doesn't freeze. In fact, if it gets much warmer than freezing, it will die. Then again, if it gets more than a degree or two cooler than freezing, it will also die.

The ice worm can survive in such cold conditions because its body contains proteins that act as a natural antifreeze. It navigates through densely packed ice crystals with the help of tiny bristles called setae that help it grip the crystals. Earthworms have setae too to help them move through soil.

It eats pollen that gets trapped in snow and algae that is specialized to live in snow and ice, as well as bacteria and other microscopic or nearly microscopic animal and plant material. In turn, lots of birds eat ice worms.

Birds also occasionally carry ice worms from one glacier or mountaintop to another by accident, which is how ice worms have spread to different areas.

The biggest ice worm species can grow nearly 2.5 inches long, or 6 centimeters, and is about 2.5 millimeters thick, but most are much smaller. The glacier ice worm can grow up to 15 mm long and is only half a millimeter thick, basically just a little thread of a worm. It only lives in glaciers. You'd think that in such an extreme environment there would only be small pockets of glacier ice worms, but researchers in 2002 estimated that the Suiattle Glacier in Washington State contained seven *billion* ice worms.

There are tall tales in the Pacific northwest about ice worms that can grow 50 feet long, or 15 meters, but those are just stories. An ice worm that big wouldn't be able to find enough to eat. Besides, there are so many mysteries surrounding the real life ice worm that there's no need to make anything up.

For instance, during the day, the ice worm hides in snow or ice to avoid the sun. It only comes to the surface from the late afternoon through morning, or on cloudy or foggy days, and usually only during the summer. But researchers aren't sure why it comes to the surface at all. Emerging from the ice is dangerous for the worm since it allows birds to find and eat it more easily, and if the air temperature is too warm or cold the worm could die. Researchers think the ice worm needs to absorb a certain amount of sunlight to augment the energy it gets from food.

Scientists don't know anything about how the ice worm reproduces. They're also not completely sure what it eats. In fact, scientists don't actually know much about the ice worm at all.

FACE SCRATCHER

During the especially hot, dry summer of 2002, stories of a small but hideous insect with spiny legs caused panic throughout the Uttar Pradesh state in India. The insect was called the muhnochwa, or face scratcher, and it supposedly came out at night. It would climb on people who were sleeping, scratching them badly with its legs and sometimes causing burns or even killing people. Some witnesses said it was the size of a football and that it glowed or sparkled with red and blue lights.

Then, in late August, someone trapped a face scratcher and took it to Lucknow University for identification.

It turned out to be a type of dune cricket, usually only found in sandy ground near riverbanks in parts of India, Pakistan, Sri Lanka, and Myanmar. It grows around 3 inches long, or almost 8 centimeters, and is yellowish-brown with sturdy legs that do indeed have spiny structures at the ends. It's nocturnal although it doesn't glow or shine.

During the day, the dune cricket lives in burrows it digs in the sandy soil, often very deep burrows since the cricket prefers damp ground. It comes out at night to hunt insects, especially grasshoppers, beetles, and crickets, including other dune crickets. Its antennae are longer than its body and the spines on its legs help it burrow and navigate the sandy soil where it lives.

So while the cricket is scary-looking, it's not dangerous to humans at all. It certainly couldn't kill anyone, and probably couldn't do more than make faint scratches that wouldn't even pierce the skin.

Unusually dry weather caused the crickets to search for moist ground, which means they were seen in areas where they were usually extremely rare. Because of its ferocious appearance, people assumed it was dangerous. Even after the insect was identified, news outlets kept reporting it as a monstrous, possibly extraterrestrial creature, which made things worse. Fortunately, it eventually turned into an urban legend sort of joke once people realized it wasn't really dangerous.

~

Venezuelan Poodle Moth

IN LATE 2008 and early 2009, a zoologist named Arthur Anker was in southeastern Venezuela in South America, and photographed a fuzzy white moth he found. He didn't know what it was so he labeled it as a poodle moth when he posted the picture online.

There are plenty of cute, fuzzy moths that sometimes get called poodle moths, such as the silkworm moth. Silkworm moths are native to Asia and are one of the few domesticated insects in the world, together with the honeybee. If you've ever had a silk shirt, that silk probably came from the domestic silkworm, which has been raised for at least 5,000 years in China and other places. The domestic silkworm moth is covered in short white hairs that make it look fuzzy.

Despite all the fluffy-looking white moths known from around the world, no one knows what species Dr. Anker's moth is. Some people have even accused Dr. Anker of making it up completely. Considering how many thousands of moths live in Venezuela, and how many new moth species are discovered every year, it's likely that the poodle moth is new to science. The trouble is that no one has seen it since. Anker wasn't on a collecting trip and he didn't realize the poodle moth might be something new to science, so he just took a picture of it and left it alone.

The best guess by entomologists who have examined the picture is that the poodle moth is a member of the genus Artace, possibly a close relation of the dot-lined white moth. The dot-lined white moth is white and fuzzy with tiny black dots on its wings. It mostly lives in the southeastern United States but there have been sightings in Colombia, which is a country in South America just west of Venezuela.

Hopefully soon a scientist can find and capture a Venezuelan poodle moth and solve the mystery once and for all. Hopefully that scientist will also take lots of pictures so we can verify that it's just as cute as it looks in its first picture.

GIANT CENTIPEDES

Centipedes have been around for some 430 million years and there are thousands of species alive today. Some of them are entirely too large.

A centipede has a flattened head with a pair of long mandibles and antennae. The body is also flattened and made up of segments, a different number of segments depending on the centipede's species, but at least 15. Each segment has a pair of legs except for the last two segments, which have no legs. Like other arthropods, the centipede has to molt its exoskeleton to grow larger. When it does, some species grow more segments and legs. Others hatch with all the segments and legs they'll ever have.

The first segment's legs project forward and end in sharp claws with venom glands. These legs are called forcipules and they actually look like pincers. No other animal has forcipules, only centipedes. The centipede uses its forcipules to capture and hold prey. The last pair of legs points backwards and sometimes look like tail stingers, but they're just modified legs that act as sensory antennae.

Different species of centipede have different numbers of legs, from only 30 to something like 300. Each pair of legs is a little longer than the pair in front of it, which helps keep the legs from bumping into each other when the centipede walks.

The centipede lives throughout the world, even in the Arctic and in deserts, which is odd because the centipede's exoskeleton doesn't have the wax-like coating that other insects and arachnids have. As a result, it needs a moist environment so it won't lose too much moisture from its body and die. It likes rotten wood, leaf litter, soil—especially soil under stones—and basements. Some centipedes have no eyes at all, many have eyes that can only sense light and dark, and some have relatively sophisticated compound eyes. Most centipedes are nocturnal.

The largest centipedes alive today belong to the genus Scolopendra. This genus includes the Amazonian giant centipede, which can grow over a foot long, or 30 centimeters. It's reddish or black with yellow bands on the legs, and lives in parts of South America and the Caribbean. It eats insects, tarantulas, frogs, birds, mice, and many other small animals. It's even been known to catch bats in midair by hanging down from cave ceilings and grabbing the bat as it flies by.

You'll often hear that the Amazonian giant centipede is the longest in the world, but this isn't actually the case. Its close relation, the Galapagos centipede, is substantially longer. It can grow 17 inches long, or 43 centimeters, and is black with red legs.

Another member of Scolopendra is the waterfall centipede, which only grows 8 inches long, or 20 centimeters, but which is amphibious. The waterfall centipede was only discovered in 2000, when entomologist George Beccaloni was on his honeymoon in Thailand. Naturally he was poking around looking for bugs, and I trust his spouse was aware that that's what he would do on his honeymoon, when he spotted a dark greenish-black centipede with long legs. It ran into the water and hid under a rock, which he knew was extremely odd behavior for a centipede. They need moisture but they avoid entering water. Beccaloni noted that the centipede was able to swim in an eel-like manner. He captured it and later determined it was a new species. Only four specimens have been found so far in various parts of South Asia. Beccaloni hypothesizes that it eats aquatic insects.

Another big centipede, the Phillip Island centipede, isn't a member of Scolopendra but is just as big. It can grow up to a foot long, or just over 30 centimeters. It lives on two small, remote islands in the Pacific that are sort of near Australia and New Zealand, but not all that near. Every year, thousands and thousands of birds nest on the islands, especially black-winged petrels, and every year, the Phillip Island centipede kills and eats several thousand of the baby chicks. There's some observational evidence that the Phillip Island centipede may even hunt in cooperative groups.

The islands have no other predators these days, although until the 1980s they were overrun with introduced pigs, goats, and rabbits that almost drove the centipede to extinction. When conservationists finally removed all the invasive species, the black-winged petrels and other birds returned and the centipede rebounded in numbers. The island ecosystem is much healthier these days.

There are stories of huge centipedes found in the depths of jungles throughout the world, centipedes longer than a grown man is tall. These are most likely tall tales, since centipedes breathe through tiny notches in their exoskeleton like other arthropods and don't have proper lungs. Arthropods can't get too big or they simply can't get enough oxygen to live.

Some of the stories of huge unknown centipedes have an unsettling ring of truth, though. There are stories from the Ozark Mountains in North America about centipedes that grow as long as 18 inches, or almost 46 centimeters.

> The biggest centipede found in the Ozarks that I have a record of was captured alive by Bent Music on Jimmies Creek in Marion County in 1860. Henry Onstott an uncle of the writer and Harvey Laughlin who was a cousin of mine kept a drugstore in Yellville and collected rare specimens of lizards, serpents, spiders, horned frogs and centipedes... Amongst the collection was the monster centipede mentioned above. [...] Brice Milum, who was a merchant at Yellville when Mr. Music brought the centipede to town, says that he assisted in the measuring of it, before it was put in the alcohol and its length was found to be 18 inches. It attracted a great deal of attention and was the largest centipede the writer ever saw. The jar with its contents was either destroyed or carried off during the heat of the war.[1]

There are large Scolopendra centipedes around the Ozarks, including *S. heros*, the Texas red-headed centipede, that can grow over 8 inches long, or

20 centimeters. Some people keep centipedes as pets and report that in captivity, the Texas red-headed centipede can grow well over 10 inches long, or 26 centimeters. I wouldn't be a bit surprised if individuals sometimes grow much longer than that.

GIANT SPIDERS

From time to time, a picture of an enormous spider pops up online and gets shared around. It's generally the size of a car at least.

You don't have to worry about spiders anywhere near that big. A spider's respiratory system isn't nearly as efficient as that of most vertebrates, so giant spiders wouldn't be able to get enough oxygen to function.

Specifically, some spiders have a tracheal system of breathing, which is the same as in most insects and other arthropods. These are breathing tubes that allow air to pass through the exoskeleton and into the body, but it's a passive process and spiders don't actually breathe in and out. Other spiders have what are called book lungs. The book lung is made up of a stack of soft plates sort of like the pages of a book. Oxygen passes through the plates and is absorbed into the blood (which, by the way, is pale blue). This is also a passive process.

In other words, any pictures of a gigantic spider (or insect, or other arthropod) are Photoshopped or made to look bigger by forced perspective. Also, spiders with wings are Photoshopped, because no spider has ever had wings, even fossil spiders all the way back to the dawn of spider history over 300 million years ago. So that's one less thing to worry about.

Stories of huge spiders are common, though, because people love to scare each other. For instance, this account from central Africa. In 1938, an

BEYOND BIGFOOT & NESSIE

English couple, Reginald and Margurite Lloyd, were driving through the jungle when what looked like a monkey or cat stepped onto the dirt road. They stopped the car so it could cross the road, at which point they saw it was a spider. It looked like a tarantula but was huge, with a legspan of up to 3 feet, or 92 centimeters. Before Reginald Lloyd could grab his camera, the spider disappeared into the undergrowth.

Supposedly, the same giant spider was reported in the 1890s by a British missionary named Arthur John Simes. Some of his men got tangled in a huge web and a pair of spiders came out and attacked them. The larger of the spiders, presumably the female, was 4 feet across, or 1.2 meters. Simes was bitten but shot one of the spiders and was able to escape. He ultimately died of the bite.

This seems less than believable, to put it gently. The largest spider that catches prey with a web is the golden silk orbweaver, but its legspan is only 5 inches across, or 12 centimeters. The biggest spiders in the world are tarantulas and other spiders that hunt actively, none of which build webs.

The Baka people from that part of central Africa are supposedly familiar with the giant spider that killed Simes. They reportedly call it the j'ba fofi and say it's rare these days, although it used to be quite common. It's venomous and spins webs between trees to catch animals, or in some accounts it only spins triplines across game trails, not actual webs. It's easy to dismiss these accounts as folklore, but there is a detail associated with the j'ba fofi that seems much more realistic. Newly hatched j'ba fofi spiders are supposed to be yellow and purple, although they turn brown as they mature. There really are some large tarantulas in Cameroon and surrounding parts of central Africa, including the giant baboon spider that can have a legspan of 8 inches, or 20 centimeters. The giant baboon spider turns a reddish-brown right before it molts its exoskeleton. It's possible that there's another big tarantula in the area, unknown to science and increasingly rare due to habitat loss, with a similar color change that inspired the story.

The largest spider known today is arguably the giant huntsman, which lives in Laos in Southeast Asia. Its legspan can be a foot across, or 30 centimeters. The goliath birdeater tarantula of northern South America has slightly shorter legs but is much heavier than the giant huntsman, weighing over 6 ounces, or 175 grams. There may be larger spiders alive today, either species unknown to science or simply unusually large individuals of known species, but they aren't anywhere near 3 feet across.

A more believable giant-spider mystery is called the up-island spider, which is supposed to be an extra-large variety of wolf spider from parts of Maine in the United States. Its legspan is supposed to be as much as 8 inches across, or 20 centimeters. Wolf spiders are common throughout the world and while they look scary, they bite people very rarely and their venom is weak, no worse than a bee sting. The wolf spider with the biggest legspan is *Hogna ingens*, with a legspan less than 5 inches across, or 12 centimeters. *Hogna ingens* lives on one island in the Madeira archipelago and is a beautiful soft grey with white stripes on its legs. It's critically endangered, but Bristol Zoo in England has a successful captive breeding program underway so it won't go extinct. The species of wolf spider most commonly found in Maine is probably *Tigrosa helluo*, but it's not very big, only a couple of inches across at most, or maybe 5 centimeters.

It's likely that the up-island spider is actually the Carolina wolf spider, which does live in Maine, although it's not very common there. It can have a legspan of 4 inches, or 10 centimeters. I can tell you from personal experience that they look a whole lot bigger if you see one in your garage or basement when you flip on the light.

TULLY MONSTER

T hree hundred million years ago, in what is now the state of Illinois in North America, a strange animal lived in the shallow sea that covered part of the area. The land that bordered this sea was swampy, with many rivers emptying into the ocean. These river waters carried dead plant materials and mud, which settled to the bottom of the ocean. When an animal died, assuming it wasn't eaten by something else, its body sank into this soft muddy mess. The bacteria in the mud produced carbon dioxide that combined with iron also present in the mud, which formed a mineral called siderite that encased the dead animal. This slowed decay long enough that an impression of the body formed in the mud, and as the centuries passed and the mud became stone, the newly formed fossil was surrounded by a protective ironstone nodule. That's why we know the Tully monster's body shape even though it was a soft-bodied animal.

An amateur fossil collector named Francis Tully discovered a fossil of the strange creature in 1955. The paleontologists he showed it to had no idea what it was or even what it might be related to. It was described in 1966 and given the name Tullimonstrum, or "Mr. Tully's monster," which is pretty much what everyone was calling it already.

The Tully monster grew up to 14 inches long, or 35 centimeters. There's evidence of light and dark banding along its body, which may indicate its body was segmented. It was shaped sort of like a slug or a leech, with eyes on stalks that jutted out sideways, although the stalks were more of a bar that grew across the top of the head. The tail end had two vertical fins, which argues that the Tully monster was probably a good swimmer. But at the front of its body was a long, thin, jointed proboscis that ended in claws or pincers lined with eight tiny tooth-like structures. It's easy to assume that the pincers acted as jaws and therefore the proboscis was a mouth on a jointed stalk, but we really don't know. It resembles nothing else known, and is so bizarre that researchers aren't sure where to place it taxonomically.

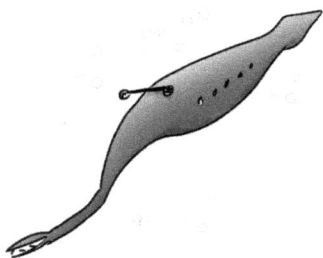

Weird as it was, it wasn't rare. Paleontologists have since found lots of Tully monster fossils in the Illinois fossil beds, known as the Mazon Creek formation. The Mazon Creek formation is also the source of highly detailed fossils of hundreds of other plant and animal species, including some that have never been found anywhere else.

Scientists have suggested any number of animal groups that the Tully monster might belong to. It might be a type of arthropod, a mollusk, a segmented worm...or it might be a vertebrate. The tiny tooth-like structures in the pincers have been analyzed and researchers think they were more similar to keratin than chitin. Keratin is a vertebrate protein while chitin is an invertebrate protein.

In 2016 a study argued that pigments in the eyes are arranged the same way as they are in vertebrates, which meant the Tully monster might have been a vertebrate. The problem is that some invertebrates also have these same pigment arrangements, notably cephalopods like octopuses. A 2019 study also looked at the chemical makeup of the fossil eyes, this time with even more advanced equipment—specifically, a synchrotron radiation lightsource, which is a type of particle accelerator. It sounds so science-y. This study suggested that the Tully monster's eyes had a different chemical makeup than the vertebrates found in the same fossil beds, which means the Tully monster probably wasn't a vertebrate after all. But it also didn't match up with known invertebrates from the same fossil beds. By that measure, it's neither a vertebrate nor an invertebrate.

Of course, it might be a deuterostome. The animals in this superphylum develop a nerve cord at some stage of life, usually as an embryo, but may not retain it into adulthood. This includes echinoderms such as sea stars and sea urchins, may include acorn worms although some scientists disagree, and tunicates like sea squirts, among others. It also includes all vertebrates. So it's certainly possible that the Tully monster might not be an actual vertebrate but might be a deuterostome, which includes the subphylum vertebrata.

One suggestion is that the Tully monster is related to a type of animal called a conodont. Technically the term conodont refers to its teeth, with the animal itself known as conodontophora, but conodont is easier to spell. We know very little about the conodont, since almost the only fossils we have of it are tiny teeth. We also have eleven body impressions, so we know it was long and skinny like an eel and grew up to 20 inches long, or 51 centimeters. We also know it had large eyes, a notochord (or primitive spine), and fins on the tail end. Conodont teeth first appear in the fossil record during the Cambrian, some 525 million years ago. They disappear entirely from the fossil record about 200 million years ago during the end-Triassic extinction event. During those 300-some million years they were around, they left so many tiny fossil teeth that they're considered an index fossil, which helps scientists determine how old a particular strata of rock is.

When I say tiny teeth, I mean *tiny*. They're microfossils usually measured in micrometers, although some of the larger ones were as much as 6 millimeters long. But they weren't teeth like modern animal teeth, and the mouth wasn't like anything we know today. This was long before jawed animals evolved some 400 million years ago.

The conodont's mouth is called a feeding apparatus by scientists and its teeth are technically known as conodont elements since they're not really teeth. There were three types of the conodont elements, meaning they had different shapes and probably different functions.

Some species of conodont may have used the elements to crush prey, but they probably weren't very strong swimmers so may have mostly eaten very small animals. Some researchers even suggest the conodont used the elements to filter plankton from the water, while others think the conodont might have been parasitic on larger animals, like the sea lamprey is. Conodonts were probably related to hagfish and lampreys and may have looked similar, although not everyone agrees with this classification. Some researchers even think conodonts might have been invertebrates.

That brings us back to the Tully monster. The Tully monster may have used its proboscis to probe for food in the mud at the bottom of the sea, but that's just a guess. Because the proboscis had a joint, it probably couldn't act as a sort of straw. The pincers may have grasped tiny prey and conveyed it to a mouth that hasn't been preserved on the specimens we have.

Another possibility is that the Tully monster was related to Anomalo-carids. Anomalocaris and its relations were strange arthropods that developed during the Cambrian. It had eyes on stalks, clawed appendages that grew from its front near the mouth, and the rear of its body was segmented with tail fins. Another Cambrian arthropod, Opabinia, had a single flexible feeding proboscis with claws at the end, five eyes on stalks, and a segmented body, so the Tully monster may have been related to it. But we don't have anything definitive yet one way or another as to what it was related to.

PART EIGHT
MYSTERY PRIMATES

Some primates look like you could buddy up with them and hang out, maybe solve some crimes. Other primates are cute. But the primates in this section are all weird in one way or another. From the sad case of Oliver the "ape man" to the Chinese ink monkey, and even one mystery lemur, let's find out more about some of our mysterious distant cousins.

CHINESE INK MONKEY

The story goes that as far back as 2000 BCE, a tiny primate known as an ink monkey was frequently the pet of scholars and scribes in China. It wasn't just a cute little pet, it was useful. It was intelligent and could be trained to prepare ink, which back in those days came in blocks and had to be ground into powder and mixed with water to the right consistency. It would turn book pages so the scholar could read hands-free, it would hand pens and other items to the scholar, and it was small enough to sleep in the scholar's brush pot or desk drawer. Such a useful little creature was highly sought after, but was supposed to have gone extinct at some point centuries ago.

According to a book of Chinese lore called *The Dragon Book*, published in English in 1938, the ink monkey was only around 5 inches long, or 13 centimeters. Its sleek fur was black and soft and it had red eyes. It was also supposed to drink any ink remaining at the end of the day as its preferred food.

Since ink in those days was frequently made with precious materials like sandalwood, crushed pearls, musk, rare herbs, and even gold, and those things are not just valuable, they're not all that nutritious, ink monkeys probably didn't actually drink ink. But was it even a real animal or just a legend?

In April of 1996, the ink monkey story got media attention when a press release from the official New China News Agency announced its rediscovery in the Wuyi Mountains of Fujian Province. The press release didn't have many details at all. It basically just reported that the animal was mouse-sized and had been found.

The smallest monkey alive today is the pygmy marmoset from South America, which is about 10 inches long, or almost 26 centimeters. But there is another animal that looks like a monkey but which is no more than about 6 inches long, or 15 centimeters, not counting its tail.

The tarsier is a nocturnal primate with huge round eyes, mouse-like ears, and sucker-like discs at the ends of its toes which it uses to climb trees. Its tail is extremely long, as are its hind legs, which helps it jump through the trees where it spends almost its whole life. While the various species of tarsier are only found on various islands of Southeast Asia today, they were once more widespread. One extinct species did live in China, but not recently. Not even remotely recently. More like 35 to 40 million years ago.

The smallest species is the pygmy tarsier, which is only found in central Sulawesi in Indonesia. It was thought extinct for decades until 2000, when it was rediscovered by local scientists. It's only about 4 inches long, or 10.5 centimeters.

The tarsier's eyes are each as big as its brain and most pictures of tarsiers taken in daylight show it squinting as though it's considering an important philosophical question. The tarsier's fur is soft, usually beige or orangey in color, and its eyes are golden. It communicates in ultrasound—sounds too high for humans to hear. It's carnivorous and mostly eats insects, but it will also eat birds, bats, and reptiles when it can catch them.

At least one imminent naturalist, Sir David Attenborough himself, suggested that a species of tarsier might easily have been living in China all along without being known to science. While it is doubtful that a tarsier could learn to prepare ink or turn book pages, it's also possible that if a

famous scholar kept one as a pet, the story of its helpfulness might have been added over the centuries.

The mystery of the ink monkey's rediscovery was cleared up by zoologist Karl Shuker. He discovered that a few weeks before the official press release, a short account of a discovery was published in the London *Times* on April 5, 1996. That report was about the discovery of a mouse-sized primate in China, sure, but not a living animal. This was a fossil discovery—specifically, a fossil jaw of a tiny proto-monkey that lived around 43 million years ago.

As Shuker concludes, the confusion probably stems from a poor English translation in the press release, leading to reporters thinking a live animal had been discovered.

That doesn't mean there wasn't once a real primate that gave rise to the Chinese ink monkey legend, whether it's a tarsier or an actual monkey or something else.

OLIVER THE "APE-MAN"

Oliver was a strange-looking chimpanzee sometimes referred to as an ape-man back in the 1970s. Oliver had been part of a traveling animal act but he never fit in with the other chimps and preferred to spend his time with humans, helping with chores. He walked fully upright at all times.

In 1976 an attorney called Michael Miller bought Oliver, mostly because Oliver just looked weird. His head was oddly shaped compared to other chimps and his jaw was smaller and more human-like in appearance. His ears were slightly pointed. The popular press found Oliver interesting and for a short while he was famous. Some claims about Oliver were that he had 47 chromosomes instead of a chimp's normal 48, that he was a mutant, that he was a hybrid between a chimp and some other primate, or even an ape-man somewhere between a human and a chimp.

Oliver had a rough life. Michael Miller sold him to a theme park in 1977 and after that Oliver was passed from theme park to theme park. Interest in Oliver died down after a while and in 1989, he was bought by a laboratory that leased out animals for testing. Oliver was never used as an experimental animal but he lived for seven years in a cage so small he could barely move, so that his muscles atrophied.

Fortunately, in 1996 Oliver finally got a break and moved to an animal sanctuary in Texas. He had a spacious territory of his own, a chimp mate called Raisin, and lived out the rest of his days in peace. He died in 2012 at the age of about 55.

When the sanctuary acquired Oliver, they had him genetically tested to see if he really was a hybrid animal. It turned out that Oliver's chromosome count was normal for a chimpanzee and that he was genetically normal in every respect. So why did he look so weird?

Mainly, it was because his teeth had all been pulled at an early age so he couldn't bite. This barbaric practice resulted in his jaw muscles being underdeveloped and his jaw bones becoming shortened. His head and ear shape were well within normal range for chimps but looked strange when combined with his poorly developed jaw. The reason he walked upright all the time was because he'd been trained to do so.

After Oliver died, the sanctuary cremated his body and spread his ashes on the grounds where he had lived peacefully for the first time in his life.

BILI APE

A Belgian army officer donated some unusual skulls to a museum in Belgium in 1898. He said the skulls were from gorillas killed in what is now the Democratic Republic of Congo. Specifically, he said the gorillas lived in a forest near the village of Bili in an area referred to as Bondo. After a museum curator examined the skulls and realized they weren't the same as other gorilla skulls, and not from an area where gorillas were known to live, the mystery ape was dubbed the Bili ape or the Bondo ape. The curator thought the Bili ape was a subspecies of gorilla.

In 1970 a mammalogist examined the skulls and determined that they were just regular old western lowland gorilla skulls. Nothing exciting. But a conservationist and photographer named Karl Ammann wasn't convinced. He decided to go out and see if he could find the Bili ape for himself, take pictures, and learn what the ape really was. In 1996, he took his cameras and went looking for gorillas.

He didn't find any but he did find a skull. It looked sort of like a gorilla skull, which has what's called a sagittal crest that runs along the top of the skull and allows the attachment of a gorilla's powerful jaw muscles. But the rest of the skull looked more like a chimpanzee's. Ammann also bought a photograph taken from a poacher's trail cam that showed what looked like huge chimps. He found great big poops and great big footprints too, larger even than a gorilla's footprint.

He had enough evidence to interest researchers, so in 2001 he and a team of scientists returned to find the Bili ape. They had no luck, partly because there was a civil war going on in the area at the time and getting around without getting killed was difficult. But they did find evidence that the apes were there, and the evidence was confusing. Gorillas build nests on the ground to sleep in, and the team did find big nests on the ground. But gorillas don't like swampy ground, plus they move around a lot and build a new nest every night. These nests were often in swampy areas and showed evidence that they were reused. Chimps prefer to sleep in trees. But while the feces the researchers collected from around the nests were big enough to be gorilla poops, they indicated the apes' diet was high in fruit, which is typical of chimps.

The team returned to the area in 2003 after the civil war ended, hoping to learn more. This time they found the Bili ape.

The first scientist to see a Bili ape was a primate behavior specialist named Shelly Williams. The whole group heard the apes in the trees close around them—and then four apes rushed out at them. Williams knew they weren't just trying to intimidate the humans, they were going to kill them. Being a primate behavior specialist means you know when you're about to die at the hands of an enraged mystery ape. But the apes caught sight of her, stopped short, and returned into the brush.

Williams also knew that the apes weren't after the humans specifically but had responded to a call made by the team's tracker, who had imitated the noise a wounded antelope makes. Imagine the scene from the apes' point of view. You're out hunting with your buddies, you hear some loud noises of animals walking through the forest. Then you hear an antelope. You and your buddies rush out, already thinking about how good that antelope is going to taste—and instead of antelopes, you see a bunch of humans. Of course you're going to beat feet, because those humans might be hunting *you*.

Williams was the only scientist in the group to get a look at the apes that day, and they confused her. They mostly looked like chimps but they were huge. A male common chimpanzee is about 5 feet tall when standing, or 1.5 meters, with females usually about a foot, or 30 centimeters, shorter. The

Bili ape was the size of a gorilla, closer to 6 feet tall, or 1.8 meters. Williams wasn't sure if she'd seen giant chimps or weird gorillas or something else entirely.

After that first sighting, the team was able to get video and photos of the Bili apes. They resemble large chimps with gorilla-like heads, and Williams thinks the females and young mostly sleep in trees while adult males sleep on the ground. They seem to live and travel in small groups, compared to chimps that usually live in troops of up to 50 members.

The locals in the area say there are two different kinds of Bili ape. The smaller ones prefer to live in trees and are known as tree-beaters. The larger ones live on the ground and are called lion-killers. The lion-killers are supposed to be immune to the poison-dart frog secretions that locals use to poison their arrow tips.

Genetic samples from dung and hair finally cleared up the mystery. Results indicate that the apes are chimpanzees, specifically a known subspecies of the common chimpanzee. Researchers think the Bili ape may look and act different since it's so isolated from other chimps and may be somewhat inbred. Bili apes encountered far from villages show very little fear of or aggression toward humans, only curiosity. Unfortunately, the chimps are under increased threat from poaching, since gold mining began in the area in 2007 and the population of humans has increased. Hopefully protections can be put into place soon so these rare chimpanzees can remain safely in their homes and can continue to be studied by researchers.

One exciting thing to remember is that the area where the Bili ape lives is still quite remote. There could very well be other animals unknown to science hidden in the forests. That's yet another reason to protect the forest and everything that lives in it. You never know what might be out there ready to be discovered.

K⊙OLAKAMBA

Aphotograph of a purported koolakamba, taken at the Yaounde Zoo in Central Cameroon in Africa, appeared in the November 1996 issue of the *Newsletter of the Internal Primate Protection League*. The ape in the picture was a male called Antoine. He had very black skin on his face but bright orange eyes with a pronounced brow ridge, something like a chimpanzee with gorilla-like features.

The first European to write about the koolakamba, in the mid-19th century, was a man called Paul DuChaillu. He was also the first European to write about several other animals, including the gorilla, and he was always eager to find more and describe them scientifically. DuChaillu was the one who gave the koolakamba its name, which was supposed to be a local name for the animal, meaning "one who says kooloo." In other words, the ape's typical call was supposed to sound like it was saying kooloo.

Chimpanzees and gorillas were well known to the local people, of course, but although they weren't "discovered" until much later, early travelers to Africa mentioned them occasionally. The first mention of both dates to about 1600. In 1773 a British merchant wrote about *three* apes he heard about from locals: the chimpanzee, the gorilla, and a third ape called the itsena.

DuChaillu thought the koolakamba was a separate species too, one that looked similar to both the gorilla and the chimpanzee. Other explorers, big game hunters, and zoologists thought it was a chimp-gorilla hybrid, which accounted for its similarity to both apes. A few thought the koolakamba was just a subspecies of chimp, while a few thought it was a subspecies of gorilla.

The argument of what precisely the koolakamba *is* is still ongoing, but no one ever denied that the koolakamba existed. After all, there were specimens, both dead and alive. In July 1873, a female chimpanzee named Mafuka was shipped to the Dresden Zoo, and she was supposed to be a koolakamba.

We have some beautifully done engravings of her face that are so detailed they might as well be photographs. Mafuka had black skin on her face, pronounced brow ridges, fairly small ears, and a gorilla-like nose. Her hair was black with a reddish tinge. She was also a big ape although she was young, measuring almost 4 feet tall, or 120 centimeters. She only lived two and a half years in captivity, unfortunately, dying in December of 1875.

Some zoologists classified Mafuka as a young gorilla, while others thought she was a chimpanzee. Others thought she was a hybrid of the two apes. In 1899 an anatomist claimed she was a koolakamba and a different species from either ape.

Other koolakamba apes have been identified after Mafuka, including one called Johanna kept by Barnum & Bailey at the end of the 19th century, but there are more recent examples. A chimpanzee colony kept at the Holloman Air Force Base in New Mexico supposedly had a koolakamba in the 1960s. An ape expert named Osman Hill studied the chimps at Holloman and published his observations in the late 1960s in a comprehensive taxonomy of the chimpanzee. Hill was convinced that the koolakamba was a subspecies of chimp, which he named *Pan troglodytes kooloo-kamba*.

But Hill's description of the koolakamba varies from DuChaillu's description. Basically the only agreements between the two is that the koolakamba has a black face—dark enough that it's usually referred to as ebony—and pronounced brow ridges.

No one seems able to agree on what the koolakamba definitively looks like. Part of the problem is that Europeans who went to Africa to kill animals and claim them as new to science asked the locals what a particular animal was, and assumed that the locals thought about animal relationships the same way Europeans do. Europeans think of animals as distinct species even if they look similar, but many people in Africa, especially hunters, and especially in the 19th century and earlier, approached animals with a different mindset. They needed to know which animals were good to eat, which animals were safe to hunt and which were dangerous and should be avoided, and so forth. They often gave different names to the same species of animal based on physical characteristics like size or color. But the Europeans didn't know this. Many of the local names reported for apes that resemble what we might call the koolakamba translate to things like "gorilla's brother" and "gorilla-like."

There are a lot of things going on here. Let's see if we can make some sense out of this confusion.

The first big question, of course, is if chimpanzees and gorillas even live in the same parts of Africa. It turns out they do, at least in a few places in western Africa. Where the territories of chimps and gorillas overlap, they generally avoid each other. It's rare that they interact at all, and extremely rare that they get in fights. Even if they were feeding in the same small area, they wouldn't need to fight because they eat different things. Gorillas mostly eat leaves and twigs, while chimps prefer fruit and meat. Also, of course, gorillas stay on the ground while chimps spend most of their time in trees.

Still, there's enough population overlap for gorillas and chimpanzees to potentially interact. That doesn't mean they hybridize, of course. While gorillas and chimpanzees do share a subfamily, they don't share a genus, which means they're not very closely related. Chimps are actually more closely related to humans than to gorillas, and we share the same subfamily with both. The less closely related two species of animal are, the less likely they will be interested in mating, the less likely that a pregnancy will result even if they do mate, and the less likely that the baby will survive even if the female does get pregnant. While it's unlikely that gorillas and chimps could or would hybridize, it's not completely out of the question. But even if it does happen, it would be an extremely rare occurrence for a chimp-gorilla hybrid to be born at all, much less live to adulthood.

We can make a check-mark next to the "hybrid ape" hypothesis, but only a very small check-mark.

Could the koolakamba be a separate species of ape entirely, something new to science? That wouldn't explain why it's generally seen in the company of chimpanzees that look like ordinary chimps, not other koolakambas. There are reports that the koolakamba is solitary or only hangs out on the edges of chimp societies, but I can't find any good sources for these claims and they may not be accurate. A rare species of ape related to the chimpanzee shouldn't be hanging out with chimps. Different species with the same dietary and environmental needs don't live in the same place. One will always outcompete the other, either driving it to extinction or into another area.

So I say no check-mark next to new species of ape.

The Bili ape is a population of chimps where the males grow especially large and look gorilla-like. Could the koolakamba actually be the same thing as a Bili ape? The Bili ape is only found in far northern Congo in the Bili Forest, which is close to central Africa, while the koolakamba is only reported from West Africa. So no check-mark for this hypothesis either.

Chimps can show a lot of variety in facial features, including skin color and ear shape and size. They also vary in overall body size, just as any animal does. I suspect the main reason that the koolakamba is so often considered a gorilla-chimpanzee hybrid is because the koolakamba's face is always described as ebony or jet black. This is uncommon in chimps, but all gorillas have dark gray or black skin.

Populations of the subspecies of chimp that lives in West Africa, the western chimpanzee, are so different from other chimps that some researchers suggest it may be a different species. These populations use spears to hunt, cool off by swimming and playing in water, are more social between tribes than other chimps, and even sometimes live in caves. They also typically live in savannas or open woodland instead of thick forest. Until recently, most observational studies of chimps in the wild have focused on the eastern chimpanzee, so researchers were shocked to learn how different the western chimp is. And the western chimpanzee is generally a little larger than eastern chimps.

It may be the case that the koolakamba isn't a separate type of animal but a western chimpanzee that shows individual differences that seem striking to us. The fact that even ape experts and local hunters can't agree on what the koolakamba actually looks like suggests that it's not a separate subspecies or even a hybrid. It's just a chimp who happens to have some facial features that look slightly more gorilla-like than other chimps. This is where I would put a nice big check-mark, pending new information.

DE LOYS' APE

In 1917, geologist François de Loys led an expedition to Venezuela and Colombia to search for oil. It was a disaster of an expedition, since not only did they not find oil, almost everyone in the expedition died. According to de Loys, in 1920 what was left of the group was camped along the Tarra River on the border between Colombia and Venezuela when two large animals appeared. De Loys said he thought they were bears at first, then realized they were apes of some kind. They were large, had reddish hair and no tails, and walked upright. The apes became aggressive toward the humans and, fearing for their lives, the geologists shot at the apes. They killed one and wounded the other, which fled.

The dead ape looked like a spider monkey, which was fairly common in the area, but it was much larger and had no tail. There was no way for the expedition to keep the body, so they propped it up on a crate with a stick under its chin to keep it upright, then took pictures. Only one of those pictures survived, since de Loys said the others were lost when a boat capsized later in the expedition.

After de Loys got home to Europe, he didn't tell anyone about the ape. He said he forgot all about it until 1929 when the anthropologist George Montandon noticed the surviving photograph in de Loys's papers. After that, De Loys wrote an article about the ape which was published in the *Illustrated London News*.

It was a sensational article, not meant to be scientific. Here's an excerpt:

> The jungle swished open, and a huge, dark, hairy body appeared out of the undergrowth, standing up clumsily, shaking with rage, grunting and roaring and panting as he came out onto us at the edge of the clearing. The sight was terrifying...
>
> The beast jumped about in a frenzy, shrieking loudly and beating frantically his hairy chest with his own fists; then he wrenched off at one snap a limb of a tree and, wielding it as a man would a bludgeon, murderously made for me. I had to shoot.

Montandon was enthusiastic about the ape. He wrote three articles for scientific journals and proposed the name *Ameranthropoides loysi* for it. But scientists were skeptical. Who was this de Loys guy and did he have any proof that the ape wasn't just a spider monkey? Did he even have proof that the photograph was taken in South America?

Quite apart from what kind of primate de Loys' ape might be, if it really is an ape, is it an ape native to South America? There are no apes native to the Americas at all, only monkeys. Chimpanzees, gorillas, and bonobos live in Africa, while orangutans, gibbons, and siamangs live in Asia. If de Loys really did find an ape new to science in South America, it radically changes what we know about ape evolution.

De Loys said he measured the animal as 157 centimeters high, which works out to about 4.5 feet. This is much larger than a spider monkey, which tops out at about 3.5 feet high, or 110 centimeters. But we have only de Loys's word to go by, and as it happens, de Loys was a known practical joker. He also didn't talk about the ape very often and seems to have only written his article at the urging of Montandon, his friend the anthropologist.

In 1962, a medical doctor, Enrique Tejera, read an article about de Loys' ape in a magazine called *The Universal*. Tejera had worked with de Loys during part of his expedition as a camp doctor, and he had firsthand knowledge about de Loys' ape. He wrote a letter to the magazine that was found and re-published in 1999 in the Venezuelan scientific magazine *Interciencia*. Here's an excerpt of the translated letter:

This monkey is a myth. I will tell you his story. Mister Montandon said that the monkey had no tail. That is for sure, but he forgot to mention something: it has no tail because it was cut off. I can assure you, gentlemen, because I saw the amputation. In 1917 I was working in a camp for oil exploration in the region of Perijá. The geologist was François de Loys and the engineer Dr. Martín Tovar Lange. De Loys was a prankster and often we laughed at his jokes. One day they gave him a monkey with an infected tail, so it was amputated. After that de Loys called him 'el hombre mono,' the monkey man.

Some time later de Loys and I entered another region of Venezuela, an area called Mene Grande. He always took his monkey along, who died some time later [in 1919]. De Loys decided to take a photo and I believe that Mr. Montandon will not deny it is the same photograph that he presented in 1929.[1]

The monkey Dr. Tejera said de Loys had been given was a white-fronted spider monkey, and that's exactly what the photo de Loys took looks like. There are no people in the photo, nothing except the crate it's sitting on to use as a size reference. You can't even see whether the animal has a tail or not.

The white-fronted spider monkey is endangered these days due to habitat loss and hunting, but in the early 20th century it was still common in Colombia, Venezuela, and other parts of northwestern South America. It's mostly black with a white belly, a long tail, and long arms and legs. That's why it's called a spider monkey, incidentally. It has long arms and legs like a spider.

The white-fronted spider monkey mostly eats fruit, but it also eats leaves, flowers, and other plant parts, and occasionally insects. Like many monkeys, its tail is somewhat prehensile and has a bare patch near the end that helps it grip branches like an extra finger. Since the spider monkey doesn't have actual thumbs on its hands like most primates, it needs that tail to help it get around in trees.

If you look closely at the photograph of de Loys' ape, you can see that the poor dead monkey does not have thumbs on its hands the way an ape would. It also looks like it has a penis, but that's actually not a penis. Female spider monkeys have an organ that retains droplets of urine and drips them out as the monkey travels around, leaving a scent trail, and which looks superficially like a penis. It's actually called a pseudo-penis and it makes it difficult for researchers to determine whether a spider monkey in the wild is male or female at first glance. It's also an organ only found in spider monkeys and a few other types of monkey, never apes.

So we can be pretty sure de Loys' ape was actually a spider monkey. But there's more going on here than a simple hoax. Here's another excerpt from de Loys's 1929 article.

> Until my discovery of the American anthropoid, we could only imagine that man migrated to these shores. But now, in the light of this discovery, it is obvious that the failure of the otherwise well established principle of evolution when it was applied to America was due only to imperfect knowledge. The gap observed in America between monkey and man has been eliminated; the discovery of the Ameranthropoid has filled it.

What is that mess of a paragraph trying to say?

Well, basically, it's promoting Montandon's theory that humans of different races evolved from different apes. We know these days that that's nonsense. All humans are genetically the same species, despite superficial physical differences like skin and hair color. Montandon thought that, for instance, people from Africa had evolved from gorillas, Asians evolved from orangutans, while people from Europe—you know, white people—were the only ones actually descended from early *Homo sapiens*.

In other words, Montandon wasn't just a terrible scientist, he was a terrible human being, because his theory was pure racism. He was delighted to learn about de Loys' ape because he decided that was the ape that Native Americans must have evolved from. Again: nonsense science, awful person, I'm glad he's dead. The French Resistance killed him during World War II.

It's possible that de Loys wasn't even trying to hoax anyone initially. He just had a pet monkey that died, took a photo as a creepy joke, and stuck the photo in his papers. It was Montandon who came across the photo and urged de Loys to write about it. It's very likely that Montandon decided to claim the animal was an ape to further his racist theory and de Loys went along with it, possibly reluctantly given how little he talked about it.

FLORES LITTLE PEOPLE

In 2003, a team of archaeologists, some from Australia and some from Indonesia, were in Indonesia to look for evidence of prehistoric human settlement. They were hoping to learn more about when humans first migrated from Asia to Australia. One of the places they searched was Liang Bua cave on the island of Flores. They found hominin remains in the cave, all right, but they were *odd*.

The first skeleton they discovered was remarkably small, only a bit more than 3.5 feet tall, or 106 centimeters, although it wasn't a child's skeleton. The skeleton was mostly complete, including the skull, and appears to be that of a woman around 30 years old. She's been nicknamed the Little Lady of Flores, or just Flo to her friends. Officially she's LB1, the type specimen for a new species of hominin, *Homo floresiensis*.

Until very recently, that statement was super controversial. In fact, there's hardly anything about the Flores remains that *aren't* controversial.

At first researchers thought the remains were only 12,000 or 13,000 years old, or 18,000 at the most. Stone tools were found in the same sediment layer where Flo was discovered, as were animal bones. The tools were small, clearly intended for hands about the size of Flo's, which argued that she was part of a small-statured species and wasn't an aberrant individual.

The following year, 2004, the team returned to the cave and found more skeletal remains. None were very complete but they were all about Flo's size. Researchers theorized that the people had evolved from a population of *Homo erectus* that arrived on the island more than three quarters of a million years before, and that they had become smaller as a type of island dwarfism. A volcanic eruption 12,000 years before had likely killed them all off, along with the pygmy elephants they hunted.

But as more research was conducted, the date of the skeletons kept getting pushed back: from 18,000 years old to 95,000 years old to 150,000 years old to 190,000 years old. Dating remains in the cave is difficult because it's been subject to flooding and partial flooding over the centuries. Currently, the skeletal remains are thought to date to 60,000 years ago and the stone tools to around 50,000 years ago.

When news of the finds was released, the press response was enthusiastic, to say the least. The skeletons were dubbed Hobbits for their small size, which made the Tolkien estate's head explode, and practically every few weeks there was another article about whether there were small people still living quietly on the island of Flores, yet to be discovered.

And, of course, there were lots of indignant scientists who were apparently personally angry that the skeletons were considered a new species of hominin instead of regular old *Homo sapiens*. Part of the issue was that only one skull has ever been found. It's definitely small, and the other skeletal remains are all correspondingly small, and the stone tools are all correspondingly small, and the skull shows a number of important differences from that of a normal human. But that doesn't necessarily mean it's not a subspecies of *Homo sapiens*, and of course that needs to be investigated. But some of the arguments got surprisingly ugly. There were even accusations that the entire find was faked. One person even suggested that the skull's teeth showed evidence of modern dental work.

Amid all this, two unfortunate things happened. First, in December 2004 an Indonesian paleoanthropologist named Teuku Jacob removed almost all the bones from Jakarta's National Research Centre of Archaeology for his own personal study for three months. When he returned them, two leg bones were missing, two jaw bones were badly damaged, and a pelvis was smashed. Then, not long after, Indonesia closed access to Liang Bua cave without explanation, although the archeological community suspected it was due to Jacob's influence, and didn't reopen it until 2007 after Jacob died.

It's important to note that Jacob was a proponent of the theory that the remains found in Liang Bua cave were microcephalic individuals of the prehistoric local population, not a new hominin species at all. He also had a history of keeping Indonesian fossils from being studied unless he specifically approved of the research.

Since then, repeated studies of the LB1 skull have suggested that *H. floresiensis* is a separate species of hominin and not *H. sapiens* with evidence of pathology. There's still plenty of research needed, of course, and hopefully more skulls will be found. But it seems clear that *H. floresiensis* isn't just a weird subspecies of *H. sapiens*.

One of the more common theories in the last few years was that *H. floresiensis* was descended from *H. erectus*, although *H. erectus* was a lot bigger and more human-like than the Flores little people. But recent studies show that *H. floresiensis* shared a common ancestor with *H. habilis* around 1.75 million years ago. *H. floresiensis* may have evolved before migrating out of Africa, or their ancestor migrated and evolved into *H. floresiensis*. Either way, they spread as far as Indonesia before dying out around 50,000 years ago.

Other hominin remains have since been found on the island. Part of a jaw and teeth were found at Mata Menge on the island of Flores, some 50 miles away from the cave, or 80 kilometers. It's around 700,000 years old and is a bit smaller than the same bones in the later skeletons. Researchers think it's an older form of *H. floresiensis*. Researchers have also discovered stone tools on the island that date to one million years old.

Possibly not coincidentally, modern humans arrived on the island about 50,000 years ago, maybe earlier, bringing with them the arts of fire, painting, making jewelry from animal bones, and killing all of our genetic cousins.

We don't know if humans deliberately killed the *H. floresiensis* people or if they just outcompeted them. It does seem pretty certain that the two hominin species coexisted on the island for at least a while. It's even possible that knowledge of the strange small people of the island has persisted in folk tales told by the Nage people of Flores. Stories about the ebu gogo have been documented for centuries. They were supposed to be little hairy people around 3 feet tall, or 91 centimeters, with broad faces and big mouths. They were fast runners with their own language and would eat anything, frequently swallowing it whole. In some stories they sometimes kidnapped human children to make the children teach them how to cook, although the children always outwitted the ebu gogo.

Supposedly, at some point, tired of their children being kidnapped and their food being stolen, villagers gave the ebu gogo palm fibers so they could make clothes. The ebu gogo took the fibers to their cave and the villagers threw a torch in after them. The fiber went up in flames and killed all of the ebu gogo.

Until the discovery of *H. floresiensis*, anthropologists assumed the stories were about macaque monkeys, but there's a genuine possibility that the ebu gogo tales are memories of the Flores little people. Articles and editorials about the link between the two have appeared in journals such as *Nature*, *Scientific American*, and *Anthropology Today*.

We still don't know for certain when *Homo floresiensis* went extinct. There may be remains that are much more recent than 50,000 years ago.

ORANG PENDEK

The island of Sumatra in South Asia is supposed to be home to a mystery ape called the orang pendek. That means "short person" in Indonesian.

The story goes that a human-like ape lives in the forests of western Sumatra. It walks on two legs, has short black, gray, or golden fur on its body with longer hair on its head, human facial features, and is a little shorter than a human. The people of the remote Sumatran villages where the orang pendek is reported say that it's enormously strong and has small feet and short legs but long arms. It mostly eats plants and will raid crops occasionally, but it also eats insects, fish, and river crabs.

Many people think the first report of the orang pendek outside of the Sumatran people is from a 14th century traveler called John of Florence. He visited China around 1342 and many other countries afterwards, including either Java or Sumatra. He reported seeing hairy men who lived on the edges of the forest, but since he also said that the hairy men planted crops and traded with the locals, it's possible he was talking about a tribe of people who lived on the outskirts of mainstream society.

The Dutch colonized Indonesia around 1820, after centuries of varying levels of control in what was known as the Dutch East Indies. Colonists reported seeing apes or strange small people in the forest, but expeditions in the 1920s and 1930s found nothing out of the ordinary.

Interest trailed off until around 1990, when a journalist named Debbie Martyr decided she was going to get to the bottom of the mystery. She had traveled to Sumatra in 1989 for a story she was writing, and while she was there she learned about the orang pendek. She spent the next fifteen-odd years interviewing witnesses and setting up camera traps, but without uncovering any proof. She did spot what she thought might be an orang pendek at least once, but got no clear photos, no remains, no conclusive footprints. Martyr states that the ape she saw had much different proportions than an orangutan, much more human-like.

Other people have searched for the orang pendek too, also without success. *National Geographic* set up camera traps between 2005 and 2009 without getting any photos of unknown apes. One expedition found some hairs that they later sent for DNA testing, but the results were inconclusive due to the hairs' poor quality and possible human contamination.

Interest in the orang pendek spiked after remains of the Flores little people were found in 2003, and after anthropologists made connections between those remains and local stories about small, mischievous people called the ebo gogo. Researchers initially thought the *Homo floresiensis* remains were only some 12,000 years old although more recent studies have pushed this back to around 50,000 years old. The island of Flores is not all that far from the island of Sumatra, so it's not out of the question that the Flores little people also lived on other islands.

Sumatra is a big island with a lot of animal and plant species found nowhere else in the world. It's certainly a big enough island to hide a population of shy apes or small human relations. But the only proof we have that the orang pendek exists, after a couple of decades of intensive searching, are a bunch of witness reports, some blurry photos and video, inconclusive plaster casts of footprints, and some ape hairs too degraded for DNA testing. If the orang pendek was a real animal, no matter how elusive, you'd think we'd at least have one good clear photo by now.

There have been hoaxes in the past. A "young orang pendek" turned out to be a dead langur monkey with its tail cut off. A video released in 2017 that purports to show a group of motorbikers who startle an orang pendek is also a hoax.

Despite the hoaxes and the lack of evidence, people are obviously seeing *something*. Witnesses include forest rangers, zoologists, hunters, and other people who know the local animals well. What could they be seeing? Let's take a look at some of the animals of Sumatra that might be mistaken for an orang pendek, at least some of the time.

The orang pendek is supposed to have small feet, about the size and shape of a human child's foot. This sounds like the tracks of the Malayan sun bear. In fact, a footprint found in 1924 that was supposed to belong to an orang pendek was identified as that of a sun bear once it was seen by an expert.

The sun bear has sleek black fur, although some are gray or reddish, and a roughly U-shaped patch of fur on its chest that varies in color from gold to almost white to reddish-orange. Its muzzle is short and its ears are small. It's the world's smallest bear, only around 3 feet long from head to tail, or 150 centimeters, and 4 feet tall when standing on its hind legs, or 1.2 meters. It has long front claws that it uses to climb trees and tear open logs to get at insect larvae, which it licks up with its long tongue. It also eats a lot of plant material, especially fruit. It's mostly nocturnal.

Orangutans also live on the island, but currently only in the northern part of the island, although they lived all over Sumatra and in Java until the end of the 19th century. The Sumatran orangutan is one of only three orangutan species in the world, and is critically endangered due to habitat loss. It's slenderer than the other species, with pale orange fur. Like the sun bear, it eats a lot of insects and fruit; unlike the sun bear, it uses tools it makes from sticks to gather insects, honey, and other foods more easily. It also uses large leaves as umbrellas. It communicates not by sound, like other orangutans, but by gestures. In short, it sounds like a pretty awesome ape. Orangutan means "forest person," if you were wondering.

Other apes and monkeys live on the island too, including several species of gibbon. Gibbons are apes, but they're not considered great apes, only lesser apes. They look more like monkeys, although they don't have tails. They're also fairly small, generally about 2 feet long, or 60 centimeters, and quite slender. They live in the treetops and swing from branch to branch, which is called brachiation. Orangutans brachiate sometimes too, but they're much heavier than gibbons and move much more slowly and cautiously. Gibbons can *move*. They also have loud voices and melodic calls.

The siamang, a type of gibbon, has a throat pouch called a gular sac which, when inflated with air, enhances the voice and helps it resonate. Family groups of siamangs sing together. The siamang has shaggy black fur and is a little larger than other gibbons, around 3 feet tall, or 91 centimeters. Its arms are long, its legs short, and it mostly eats plants, especially fruit, although it also eats insects. Like orangutans and many other animals on Sumatra, it's threatened by habitat loss.

There's one thing that the sun bear, the orangutan, and the siamang all sometimes do that is suggestive of the orang pendek. All three sometimes walk on their hind legs. Bears usually only stand on their hind legs to get a better view of something, but they can certainly walk on their hind legs if they want to. While the orangutans of Sumatra spend most of their time in trees, since the Sumatran tiger likes to eat orangutans, it can and does come down to the ground sometimes. Males in particular sometimes walk upright for short distances. Siamangs walk upright along branches and occasionally on the ground, usually with their long arms held above their heads for balance, but not always.

So we have three animals that when seen clearly, really don't look much like the orang pendek is supposed to look. None have especially human faces, although I've just spent half an hour looking at pictures of orangutans and siamangs, and those are some handsome apes. But they all have a number of features that sound like orang pendek features. They're all the right size, they can all walk upright, their fur is black, gray, or golden, and they eat the things the orang pendek is supposed to eat.

At least some orang pendek sightings are mistaken identity of these three animals, I guarantee it. Even the most knowledgeable zoologist or forest ranger can make a mistake, especially if they only catch a glimpse of the animal or see it in poor light. And, as I often point out, people tend to see what they expect to see. In a 1993 article, Debbie Martyr herself says that when she first started studying the orang pendek sightings, people in Sumatra laughed at the thought that the orang pendek was a real animal. Now they're much more open to the possibility that there may be a mystery ape in the forest.

In other words, as the legend becomes more and more popular, more and more people report seeing a mystery animal that fits the orang pendek's description. And yet, there is no more proof now than there was in 1925 of the animal's existence.

That doesn't mean there isn't an unknown ape living in Sumatra, of course. I just don't think that's what people are seeing. It would be fantastic if the orang pendek did turn out to be a real animal. It would focus more attention on the loss of rainforest and other habitats in Sumatra, and would probably bring more tourists to the island, which would help the local economy. But until someone actually finds a body or captures a live orang pendek, we have to remain skeptical.

MANDE BURUNG

T he mande burung is supposed to be a giant ape-like animal, as much as 8 or 10 feet tall, or up to 3 meters, with black hair. It lives in the remote forests of northeast India, specifically in Meghalaya.

The mande burung has long been a creature of folklore in the area, until November 1995 when someone saw one. Interest in the mande burung has increased steadily ever since and cryptozoologists from India and other parts of the world have mounted repeated expeditions to look for it. They report finding footprints up to 15 inches long, or 38 centimeters, hair from unidentified animals, and nests made from leaves and grass. But there are no photographs of the animals, no mande burung feces, no dead bodies, and very few sightings, all of them within the last few decades and some of them decidedly questionable.

It's certainly possible that there's a mystery animal living in the area. Meghalaya is heavily forested outside of the cities and farmland. Some areas of forest are considered sacred, so they've never been logged, no one's ever lived there, and no one hunts there. As a result, these sacred forests contain some of the richest habitats in all of Asia, including plants and animals that live nowhere else. It's pretty much guaranteed that there are animals living in Meghalaya that are unknown to science.

But while Meghalaya is primarily an agricultural region, tourism is becoming more and more important. A 2007 press release even talks about how the mande burung legend will bring more tourists to the area, and that a local group had started offering tours for people looking for the mande burung. That doesn't mean the sightings aren't genuine—I think most of them are—but people see what they expect to see. The more that people talk about the mande burung, the more likely people will think of it when they see a large animal they can't identify. And there are lots of big animals living in the forests of Meghalaya, including an endangered species of gibbon, four species of macaque, and three species of bears. Any of these might resemble a Bigfoot type of creature if seen in low light or poor conditions.

In 2001, a hair found in what was called a "cedar tree root den" was DNA tested. Bear and human DNA was ruled out, and the DNA results didn't match any known animals. But a follow-up test in 2008 gave a result that was just as surprising to scientists: the hair belonged to a Himalayan goral, a bovid that wasn't known to live in the area until the genetic results came in. The goral is a small antelope-like animal with short horns that lives in the southern slopes of the Himayalas. It's dark gray or gray-brown in color with a darker eel stripe along the spine.

Generally, websites that like to talk about Bigfoots mention the first DNA test but don't mention the follow-up, but the discovery of Himalayan goral hairs in Meghalaya is exciting. Who knows what else might be hiding in the forests too?

TRATRATRATRA

Tratratratra is the actual name of an animal that was supposedly common in Madagascar when the Malagasy people settled there around 2,000 years ago. It was described as a lemur about the size of a calf with a human face but hands more like a monkey's. Supposedly it still lives on Madagascar in remote, hard-to-reach areas.

Madagascar is a big island off the coast of East Africa, with smaller islands around it. It's been isolated from both Africa and Asia for 88 million years, so many of its plants and animals are found nowhere else on earth. Lemurs are one example. There are over 100 known species and subspecies of lemur on Madagascar, but lemurs are found nowhere else in the world. Even more species of lemur have gone extinct since humans settled on the island, including one that might be the tratratratra.

If you've seen the movie *Madagascar*, you have a pretty good idea of what a lemur looks like, although you may overestimate the amount of dancing they do.

Technically the lemur is a primate, although it doesn't look much like other primates at first glance. Different species can look radically different, of course, but in general they're long-bodied animals with long tails and monkey-like hands and feet with nails instead of claws. They're mostly social animals who eat plants and fruit, although some eat insects, arthropods, and other small animals. Most lemur societies are female-led. All are endangered due to habitat loss, poaching, and the illegal pet trade.

While we tend to think of apes and monkeys when we hear the word primate, the primate order contains many other types of animal. Lemurs belong to the Strepsirrhini suborder, which includes bushbabies, pottos, and lorises. Apes and monkeys belong to the Haplorhini suborder, along with tarsiers. Researchers think that the ancestors of lemurs migrated to Madagascar from Africa about 50 million years ago on rafts of vegetation. This sounds ridiculous since Madagascar is more than 300 miles, or 500 kilometers, away from Africa at its closest point, and the prevailing winds and ocean currents push floating logs and other vegetation away from the island. But 60 million years ago the currents flowed the other way. By 20 million years ago, continental drift had pushed Africa and Madagascar farther north so that the currents changed to what they are now, which helped isolate the island even further.

The smallest lemur species is the mouse lemur, which is only 11 inches long including its tail, or 27 centimeters. The largest is the indri, which is a black and white animal with long legs but no tail, which grows to almost 2.5 feet long, or 72 centimeters. In other words, even the biggest lemur alive today isn't all that big. But that didn't used to be the case. When humans first settled on the island, there were three kinds of giant lemurs. Let's take a quick look at them.

Monkey lemurs went extinct around 1,500 years ago and probably spent most of their time on the ground. They weren't huge, probably not any bigger than the indri. We don't have very many monkey lemur remains so we don't know much about it, but researchers think it primarily ate seeds, although it might have also eaten grass and leaves. Its limbs were short and powerful with short hands and feet. It had a heavy skull with big molars for grinding plant material. It probably went extinct mostly due to competition with introduced livestock like pigs.

Koala lemurs were bigger than the indri, up to 5 feet long, or 1.5 meters, and went extinct around 1,000 years ago. Incidentally, 1,000 years ago there was a terrible drought in Madagascar that caused crops to fail, lakes to dry up, and wildfires to start, and which contributed to many species going extinct. The koala lemur was shaped more like a koala than a lemur. It lived its whole life in treetops, eating leaves, and it had some weird features for a primate. Its eyes were on the sides of its head like a rabbit's or a horse's, instead of in the front of its head like all other primates. Its snout was long and tapered, but it had a big nasal area that probably indicates an enlarged upper lip, maybe even partially prehensile, that helped it gather leaves. It was also heavier than all other lemurs, and some of the remains we have show evidence that they were butchered by humans to eat.

Finally, sloth lemurs probably ate plants, fruit, and nuts, and some species may have hung from branches the way sloths do. Instead of big claws for climbing, sloth lemurs had long fingers. There were a number of species, so let's look at a few of them. Don't worry, we're getting closer and closer to the tratratratra.

Archaeoindris was the biggest lemur that we know of. It was the size of a gorilla, maybe even a little bigger, which would make it one of the largest primates that ever lived. Its skull was large and heavy, and it probably ate leaves. We don't have any hand or feet bones so we don't know if it had adaptations for climbing trees or for walking on the ground. It was already rare when humans first came to Madagascar and went extinct shortly afterwards.

Now we're up to *Palaeopropithecus ingens*, and this may be our tratratratra. It wasn't as big as Archaeoindris but it was still much bigger and heavier than the modern indri. It ate leaves, nuts, and seeds and probably spent a lot of time in the trees. Its arms and legs were powerful and it had long fingers and toes, which it used to hang from branches like a sloth. It had other adaptations, like curved arm and leg bones, that show it was adept at climbing on, hanging from, and brachiating through tree branches. Even the hands were curved so that its fingers were more like hooks. But it probably wasn't a very fast mover, so was easily hunted by humans and would have provided a lot of meat.

According to Admiral Etienne de Flacourt in his 1658 history of Madagascar, the tratratratra had a round head and human-like face, but hands and feet like a monkey. Its hair was curly and it had a short tail. He also mentioned that it was a solitary animal and that locals were afraid of it.

Palaeopropithecus ingens probably had a short tail like the still-living indri, which is a close relative. It probably had rounded ears like the indri too. It may have had wavy or curly hair too, again like the indri.

The tratratratra's name probably came from its call, which might have been an alarm bark or a chattering sound. Many local lemur names do come from the animals' calls. Some researchers consider the tratratratra's name to be a fossil sound, and one of the very few we have.

We don't know exactly when Palaeopropithecus went extinct. It might have been as recently as 400 years ago, maybe even more recently. There have only been a few modern-day sightings of an animal that might be the tratratratra. A French forester in the 1930s saw what he claimed was a gorilla-like lemur with a human-like face, 4 feet tall, or 1.2 meters, but that's about it. Whatever the tratratratra might be, it's probably extinct by now—but maybe it's hanging on in the forested hills north of Tulear, where the last fragments of Madagascar's original forests remain.

There is an interesting Malagasy tradition reported in 2003 about an ogre with the face of a human, which was helpless on smooth rocks. Since Palaeopropithecus was so well adapted to living in trees, like modern sloths it probably couldn't walk on the ground very well. Even if the tratratratra is extinct, its memory lives on in modern culture.

PART NINE
DRAGONS AND DINOSAURS

Reptiles have fascinated and terrified people literally forever. In this section we'll learn about some mystery reptiles, including Africa's emela-ntouka, the sirrush of the Middle East, and a little lizard with a startling ability that gives it the name lightbulb lizard.

MINI REX

Not all dinosaurs were enormous like sauropods and tyrannosaurs. Many were quite small even by modern standards, chicken or turkey sized animals. Even though non-avian dinosaurs went extinct after a massive asteroid strike 66 million years ago, occasionally someone spots what they think is a little dinosaur running along on its hind legs.

Many reports come from the American southwest, especially Colorado, Arizona, and Texas. For instance, in the late 1960s two teenaged brothers were looking for arrowheads near their home in Dove Creek, Colorado when they were startled by an animal running away from them at high speed. The boys said it looked like a miniature dinosaur, only about 14 inches tall, or 35 centimeters. It was kicking up so much dust as it ran on its hind legs that the boys had trouble making out details. They did note that it seemed to be brown and possibly had a row of spines running down its back, maybe even two rows of spines, similar to an iguana's. It had long hind legs and shorter front legs that it held out in front of it as it ran.

The animal left behind three-toed footprints that the boys followed until they disappeared into some brush. The boys were familiar with turkey footprints but these were different, with the toes closer together and no rear-pointing toe prints.

In April 1996, in Cortez, Colorado, a woman saw an animal run past her house on its hind legs, seemingly from a nearby pond. It was greenish-gray and stood about 3.5 feet tall, or about a meter. It had a long neck and long, tapering tail. Its hind legs had muscular thighs but were thinner below the hock joint. She didn't notice its front legs.

One night in July 2001, a woman and her grown daughter were driving near Yellow Jacket, Colorado when they noticed an animal at the edge of the road. At first the driver thought it was a small deer and slammed on the brakes so she wouldn't hit it, but when it darted across the road both women were shocked to see what looked like a small dinosaur pass through the headlight beams of the car. They reported it was about 3 feet tall, or 91 centimeters, and that it had no feathers or fur. Its legs were thin and long while its arms were tiny and held out in front of its body. It had a slender neck, a small head, and a long, tapering tail.

The witnesses in both the 1996 sighting and the 2001 sighting noted that the animal they saw ran gracefully. They also all agreed that the skin appeared smooth.

Lots of dinosaurs used to walk on their hind legs, but the reptiles living today are all four-footed. There are a few lizards that run on their hind legs occasionally, though, and one of them lives in the American southwest. The collared lizard, also called the mountain boomer, will run on its hind legs to escape predators. Females are usually light brown while males have a blue-green body and light brown head. The name collared lizard comes from the two black stripes both males and females show around their necks, with a white stripe in between. During breeding season, in early summer, females also have orange spots along their sides.

The collared lizard can run up to 16 miles an hour, or 26 kilometers per hour, for short bursts on its hind legs. It uses its long tail for balance as it runs, and its hind legs are three times the length of its front legs. This makes it a good jumper too. It mostly eats insects but will occasionally eat berries, small snakes, and even other lizards. It hibernates in winter in rock crevices.

While the teenaged boys probably saw a collared lizard in the 1960s, the other two sightings we just covered sound much different. The collared lizard typically only grows up to 14 inches long, or 35 centimeters, including its long tail.

A few other lizards are known to run on their hind legs, such as the basilisk that lives in rainforests of Central and South America. It's famous for its ability to run across water on its hind legs. It's much larger than the collared lizard, up to 2.5 feet long, or 76 centimeters, including its very long tail. It holds its front legs out to its sides when running on its hind legs, and the toes on its hind feet have flaps of skin that help stop it from sinking in water. It has a crest on its head, and the male also has crests on his back and tail. It can be brown or green in color.

The basilisk is sometimes kept as an exotic pet. In 1981 in New Kensington, Pennsylvania, four boys playing along some railroad tracks saw a green lizard that they thought was a baby dinosaur. It was 2 feet long, or 61 centimeters, and had a crest and an extremely long tail. It ran away on its hind legs but one of the boys, who was 11 years old, managed to catch it. It startled him by squealing and he dropped it again, and this time it got away. It sounds like an escaped pet basilisk.

But let's go back to our miniature dinosaur sightings from 1996 and 2001, the ones of dinosaur-like animals running gracefully on their hind legs with a long neck and long tail. These don't sound like lizards at all. When lizards run on their hind legs, they don't look much like how we imagine a tiny raptor dinosaur would look. They appear awkward while running, with their arms sticking out and their heads pointing more or less upward. While all the lizards known that can run on their hind legs have long tails, they all have relatively short necks.

There's another type of animal that's closely related to the dinosaurs, though, and every single one walks on its hind legs. That's right: birds! Even penguins, which are really strange-looking birds, walk on their hind legs only. All the birds alive today are descended from dinosaurs whose front legs evolved for flight. Even flightless birds are well adapted to walk on two legs.

Let's look at the details of those two sightings again. Both were of animals estimated as about three feet tall or a little taller, or up to about a meter, with long neck, small head, long tapering tail held above the ground, and long, strong legs that were nevertheless thin. Both also appeared smooth. In one of the sightings, the front legs were tiny and held forward; in the other, the witness didn't notice the front legs.

My suggestion is that in these two sightings, at least, the witnesses saw a particular kind of bird, a wild turkey. That may sound ridiculous if you're

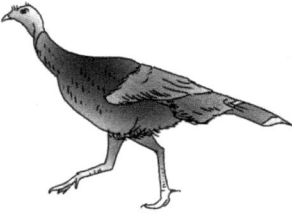

thinking of a male turkey displaying his feathers, but most of the time turkeys don't look round and poofy. Most of the time, in fact, the wild turkey's feathers are sleek and its tail is an ordinary-looking long, skinny bird tail instead of a dramatic fan. Its feathers are mostly brown and black, the upper part of its long neck is bare of feathers, as is its small head, and its legs are long and strong but relatively thin. It also typically stands 3 to 3.5 feet tall, or up to about a meter, although some big males can stand over 4 feet tall, or 1.2 meters. As for the front legs seen by witnesses in 2001, a full-grown male turkey has a tuft of long, hair-like feathers growing from the middle of his breast, called a beard. It sticks out from the rest of the feathers and might look like tiny arms if you were already convinced you were seeing a dinosaur instead of a bird.

That's not to say that all little living dinosaur sightings are of turkeys, of course, but some of them probably are. The wild turkey lives throughout much of the United States, including most of Colorado. Since birds are the closest animals we have to dinosaurs these days, though, that's still pretty neat.

BURRUNJOR

Dinosaurs once lived in what is now Australia, just as they lived throughout the rest of the world. Similar to the southwestern United States reports of little living dinosaurs, some people in northern Australia report seeing living dinosaurs running around on their hind legs—but these dinosaurs aren't so little.

The burrunjor, as it's called, is often described as looking like a *Tyrannosaurus rex*. Mostly, though, people don't actually see it. Instead, they hear roaring or bellowing and later see the tracks of a large animal that was walking on its hind legs.

One Australian dinosaur that people mention when trying to solve the mystery of the burrunjor is Muttaburrasaurus. It lived around 105 million years ago and was a herbivore that grew up to 26 feet long, or 8 meters. It walked on its hind legs and had a big bump on the top of its muzzle that made its head shape unusual. No one's sure what the bump was for, but some scientists speculate it might have been a resonant chamber so the animal could produce loud calls to attract a mate. Other scientists think it might have just been for display. Or, of course, it might have been both—or something else entirely. None of the Australian dinosaur sightings mention a big bump on the dinosaur's nose.

In fact, the stories are all pretty light on details...and they all trace back to a single source.

"Burrunjor" is supposed to be a word used by ancient Aboriginal people to describe a monstrous dinosaur-like creature that eats kangaroos. But in actuality, Burrunjor is the name of a trickster demigod in the local Arnhem Aboriginal tradition and has nothing to do with reptiles or monsters. The Aboriginal rock art supposedly depicting a dinosaur-like creature doesn't resemble other rock art in the region and isn't recognized by researchers or Aboriginal people as being authentic.

All the stories of burrunjor sightings are reported by one person, an Australian paranormal writer named Rex Gilroy. The stories Gilroy writes about don't appear in newspapers of the time. Oh, and Gilroy was the one who "discovered" the rock art of a supposed dinosaur.

Gilroy's burrunjor is probably a hoax, but there is a big lizard in Australia that sometimes stands on its hind legs. Monitor lizards live throughout Australia and are often called goannas. The largest Australian species can grow over 8 feet long, or 2.5 meters. All monitor lizards, including the Komodo dragon that lives in Indonesia, can stand on their hind legs. It does so to get a better look at the surrounding area. It uses its tail as a prop to keep it stable and can't actually walk on its hind legs, but an 8-foot lizard standing on its hind legs might look like a dinosaur from a distance.

An even bigger monitor lizard, called Megalania, lived in Australia until at least 40,000 years ago and maybe much more recently. Megalania is considered the largest terrestrial lizard known. Dinosaurs weren't lizards and crocodilians aren't either, but monitor lizards are. We don't have any complete fossils of Megalania but recent estimates of its total length, including its tail, range from 23 to 26 feet long, or 7 to 8 meters. This is more than twice the length of the Komodo dragon, the largest lizard alive today and a close relation. It was probably venomous.

It's too bad Megalania is extinct. If it was still alive today, it would definitely inspire "dinosaur" sightings and no one would have to invent those sightings in hopes of selling more books.

KAWEKAWEAU

M ost geckos are pretty small, no bigger than the length of your hand or thereabouts, but Delcourt's giant gecko is a whole lot bigger. It's about 2 feet long, or 61 centimeters. It's also extinct —at least, as far as we know. And until 1986, scientists didn't believe it had ever existed.

The Maori people in New Zealand have local lore about a big lizard called the kawekaweau. The legends were known to Europeans as early as 1777 when Captain Cook interviewed the Maori and collected stories about the the animal. In 1871 two live specimens were collected, but before any scientists could examine them, one was killed and eaten by a cat and the other escaped. In 1873 a Maori chief told a visiting biologist that he had killed a kawekaweau in 1870, and described it as "about two feet long and as thick as a man's wrist; colour brown, striped longitudinally with dull red." That was the last known sighting of the animal. Over the next century, people mostly just forgot about it. Those scientists who did read about it assumed it wasn't real.

In 1979 a herpetologist named Alain Delcourt, working in the Marseilles Natural History Museum in France, noticed a big taxidermied lizard in storage and wondered what it was. It wasn't labeled and he didn't recognize it, surprising since it was brown with red longitudinal stripes and the biggest gecko he'd ever seen. He sent photos to several reptile experts and they didn't know what it was either.

No one knew anything about the stuffed specimen, including where it was caught. At first researchers thought it might be from New Caledonia since a lot of the museum's other specimens were collected from the Pacific Islands. None of the specimens donated between 1833 and 1869 had any documentation, so it seemed probable the giant gecko was donated during that time and probably collected not long before.

A gecko specialist named Aaron Bauer traveled to New Zealand in 1984 to do a more careful study of the local geckos, since the giant gecko resembles some smaller species found in New Zealand. While he was there, Bauer learned about the kawekaweau and realized the Maori had been familiar with the giant gecko long before any scientists started poking around. In 1986 it was described as a new species.

These days, scientists acknowledge that the kawekaweau was the same animal as Delcourt's giant gecko. Further investigations revealed that a few unidentified subfossil bones found in New Zealand caves actually belonged to the giant gecko.

The only way this could be a better story is if the kawekaweau was found alive and well in remote areas of New Zealand. It's not likely but there are a few reported sightings, so maybe one day a lucky herpetologist will make the discovery of a lifetime.

～

Wee Waa Monster

In 1950, a man named George Gray started hearing bellowing from the gum swamp near his house. His adult son, Ted, heard it too and reported it was loud enough to hear half a mile away, or about 800 meters.

The bellowing sounded like a crocodile. The problem is, the Grays lived just outside of the small town of Wee Waa in New South Wales, Australia. Very, very rarely a saltwater crocodile wanders down the coast as far as New South Wales, but Wee Waa is a five or six hour drive inland. It's not crocodile country. It gets cold in winter and it's surrounded by farmland.

Ted Gray found tracks in the swamp where a large animal had flattened tall grass, although the footprints didn't have claw marks. He estimated that if the animal was a crocodile, it was only about 6 feet long, or 1.8 meters, although he did note that none of the trails seemed to lead to the water. The bellowing mostly occurred in early morning and around sundown and continued for several months.

Attempts to search the wetlands were unsuccessful, mostly because of heavy rain that spring and summer. The water was ordinarily no more than 3 feet deep, or about 91 centimeters, but was much deeper that year. Hundreds of people came to listen to the bellowing, though, and one constable perched in a tree for most of a day hoping to shoot the animal. He never saw it.

Seven years earlier, during World War II, an armored division of troops stayed in the area. The troops had been based in northwest Australia for a time and supposedly brought some baby crocodiles with them as mascots. The soldiers were told by their officers to kill the crocodiles, but newspapers in 1950 suggested they'd been released instead and that one had found its way to George Gray's swamp. That doesn't explain how it survived seven winters or why no one had seen or heard it until then.

Then again, one man named Ross Tuckey claimed he'd seen a crocodile several years before about 20 miles away, or 32 kilometers. He'd first thought it was a goanna but when his dog rushed at it, it bowled the dog away with its tail and slid into a dammed stream. Tuckey watched it float in the water with only its nostrils visible.

The Wee Waa monster was never identified.

EMELA-NTOUKA

"Emela-ntouka" is usually translated as "elephant killer" or, more poetically, "killer of elephants." Then again, it's also sometimes translated as "eater of the tops of palms." It's supposed to be the size of an elephant with hairless gray or brown skin, but with a long, heavy tail like a crocodile's and a single sharp horn on its nose. The horn is supposed to look like ivory, like an elephant's tusk. Even though it supposedly eats only plants, if any large animal enters its lake, the emela-ntouka will stab it in the belly with its nose-horn, disemboweling it.

There are no photographs of the emela-ntouka, not even bad ones. There are only a few reported sightings. In the early 1930s one was reportedly killed, but it's a friend-of-a-friend report that can't be verified. In late 1966 a photographer got pictures of some footprints along a riverbank that looked like a rhinoceros's footprints, but cryptozoologists claimed that there were enough slight differences that the footprints weren't from a rhino. And that appears to be it.

The only proof we have that the emela-ntouka is a thing at all and not a hoax, no matter whether that thing is a real animal or a folktale, are two wooden carvings and two drawings by a French artist who lived in the Congo in the late 1980s and early 1990s.

Let's start with the drawings, which are by the artist Jean Claude

Thibault. He made two drawings of the emela-ntouka, although both were misidentified when published. One drawing was published in a Congolese newspaper, date unknown, and the other in a 1996 calendar produced by the Worldwide Fund for Nature to raise funds for a project. The calendar was only sold in the Central African Republic, or CAR, and Thibault contributed at least four drawings. The calendar's theme was that of mythological creatures of the area.

Shortly after Thibault's death in 2008 or 2009, the four drawings from that calendar were exhibited in the CAR, and in 2012 or shortly thereafter the four drawings were made into cards for a local gift shop.

In 2005 a French cryptozoologist visiting Cameroon saw a wooden carving in a village and bought it. As you can probably guess, it was a carving of the emela-ntouka. Later he also mentioned that he had seen a second carving in a different location in Cameroon and made by a different artist, but he didn't buy that one, just took a photo.

The carvings greatly resemble the drawing of the emela-ntouka that appeared in the 1996 calendar, down to the pose and details like the ears. There's no account of the emela-ntouka's ears in the earliest reports of the animal, but Thibault depicted it with small elephant-like ears. Both carvings also feature small elephant-like ears.

Zoologist Karl Shuker thinks the carvings can't be influenced by the drawings because they're so different from the drawings, and because the calendar was only distributed in one small area. But the carvings are nearly identical to the drawings, and the differences come from the medium and the skill levels of the carvers. Shuker specifically mentions the way the tail in both carvings snugs up next to the legs, whereas the tail in the calendar's drawing was curved farther away from the body. But when you carve wood, you have to adjust for how much material you have. Both carvings were probably made from a block of wood that didn't allow for tails sticking out very far from the body. Not only that, if you try to carve an appendage like a tail that sticks out, it's very easy for it to break off as you work. If you keep it closer to the body and attached in several places, such as where it touches the legs as in these carvings, it's less likely to break and the whole piece is stronger.

Both carvings were found by a cryptozoologist only nine years after the calendar appeared in the CAR, which is directly east of Cameroon. We don't know when or where the carvings were made. They might have been made in the CAR in 1996 or just after, but were bought and sold several times until they ended up in Cameroon a few years later.

Let's set all that aside, though. If the emela-ntouka of folklore is based on a real animal, what might that real animal be?

It's probably not based on a rhinoceros for several reasons. Rhinos aren't aquatic, they don't have long heavy tails, and their horns are made of keratin, not ivory. In fact, ivory horns aren't a thing. Ivory is teeth, whether it's from an elephant, a walrus, or some other animal. But maybe it's supposed to be a true horn—that is, a bone that grows from the skull and is covered in a keratin sheath like a cow or antelope horn. In that case, it's even more confusing because no known living animal has a nose horn like that.

A mammal called Arsinoitherium was distantly related to the elephant and lived in northern Africa, although it wasn't as big as an actual elephant. It stood around 6 feet tall at the shoulder, or about 2 meters, and was probably at least semi-aquatic. It had a pair of huge rhinoceros-like horns on its nose rather than a single horn, but they were side by side and probably covered in a keratin sheath. But Arsinoitherium had an elephant-like tail and went extinct around 30 million years ago, long before humans evolved.

Another suggestion is that the emela-ntouka is a surviving species of ceratopsian, related to the well-known Triceratops. Different ceratopsian species had different numbers of horns, but the one typically suggested by cryptozoologists is Centrosaurus. It had a single horn at the end of its nose and a frill on the back of its skull, as well as a typical long, thick dinosaur tail. It probably grew around 18 feet long, or 5.5 meters. But it doesn't fit the emela-ntouka's description very well other than the tail and the nose horn. Centrosaurus's head frill was large and elaborate, as were those of most ceratopsian species. Centrosaurus also wasn't aquatic.

You can, of course, argue that a descendant of Centrosaurus might have evolved a small or absent head frill and become an aquatic or semi-aquatic animal, since Centrosaurus itself went extinct around 75 million years ago. But even in the unlikely event that Centrosaurus's descendants survived the extinction event that killed off all the non-avian dinosaurs, how did it also migrate from North America to Africa without leaving any traces in the fossil record? There are no ceratopsian fossils known from Africa, only a few from China, and all the rest are found in North America.

It's vanishingly unlikely that the emela-ntouka is a surviving ceratopsid. But what about Arsinoitherium? It's not likely to have survived either, but there is another reason why it might have inspired the emela-ntouka story. It's possible that the skull of an Arsinoitherium was occasionally found in the olden days, weathered out of the surrounding rock. It still doesn't explain why the emela-ntouka is described as only having one horn, but if the skull was found by itself, that would explain why the tail is supposed to be so different from the real Arsinoitherium's tail. No one knew what it was supposed to look like.

My suggestion, though, is less exciting. I think the emela-ntouka a creature of folklore and never intended to be a real animal. But that's not as much fun as imagining a living dinosaur wallowing around in a remote African swamp.

SIRRUSH

The sirrush is a word from ancient Sumerian, but it's actually not the right term for this animal. The correct term is mush-khush-shu (mušḫuššu), which means something like "the splendor serpent," but sirrush is way easier to spell. We'll go with sirrush, but be aware that that word is due to a mistranslation a hundred years ago and scholars don't actually use it anymore.

The sirrush is found throughout ancient Mesopotamian mythology. It usually looks like a snake-like animal with the front legs of a lion and the hind legs of an eagle. It's sometimes depicted with small wings and a crest of some kind, sometimes horns and sometimes frills or even a little crown. And it goes back a long, long time, appearing in ancient Sumerian art some 4,000 years ago.

Mesopotamia refers to a region in western Asia and the Middle East, basically between the Euphrates and Tigris rivers. These days the countries of Iraq and Kuwait, parts of Turkey and Syria, and a little sliver of Iran are all within what was once called Mesopotamia. It's part of what's sometimes referred to as the Fertile Crescent in the Middle East. The known history of this region goes back 5,000 years in written history, but people have lived there much, much longer. Some 50,000 years ago humans migrated into the area from Africa, found it a really nice place to live, and settled there.

Parts of it are marshy but it's overall a semi-arid climate with desert to the north. People developed agriculture in the Fertile Crescent, including irrigation, but many cultures also specialized in fishing or nomadic grazing of animals they domesticated, including sheep, goats, and camels. As the centuries passed, the cultures of the area became more and more sophisticated, with big cities, elaborate trade routes, and stupendous artwork.

That includes the Ishtar Gate, which was one of the entrances to Babylon, the capital city of the kingdom of Babylonia. The city grew along the banks of the Euphrates River until it was one of the largest cities in the world by about 1770 BCE. Probably a quarter million people lived there in its heyday around the 6th century BCE, but it was a huge and important city for hundreds of years. It's located in what is now Iraq not far from Baghdad.

Babylon is actually the source of the Tower of Babel story in the book of Genesis. In that story, people decided to build a tower high enough to touch heaven, but God didn't like that and caused the workers to all speak different languages, then scattered them across the world. But that story may have grown from earlier stories from Mesopotamia, such as a Sumerian myth where a king asks the god Enki to restore a single language to all the people building an enormous ziggurat so the workers could communicate more easily.

Babylon means "gate of the gods," and it did have many splendid gates in the massive walls surrounding the city. The ancient Greek historian Herodotus reported there were a hundred of these gates. One of these was the Ishtar Gate, built around 575 BCE. This wasn't like a garden gate but an imposing and important entry point to the city. For one thing, it was the starting point of a half-mile religious procession held at the new year, which was celebrated at the spring equinox. The gate was dedicated to the goddess Ishtar and was more than 38 feet high, or 12 meters, and faced with glazed bricks. The background bricks were blue, with decorative motifs in orange and white, and there were rows of bas-relief lions, bulls, and sirrushes.

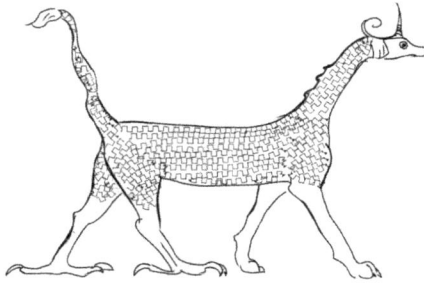

Bas relief from the Ishtar Gate

The sirrush was considered a sacred animal of both Babylon and its patron god, Marduk. It's sometimes called a dragon in English, but from artwork that shows both Marduk and a sirrush, the sirrush was small, maybe the size of a big dog.

The question, of course, is whether the sirrush was based on a real animal or if it was an entirely mythical creature.

Every culture has stories that impart useful information—warnings, history lessons, and so forth. Every culture has monsters and mythological creatures of various kinds. That doesn't mean those animals were ever thought of as real animals, although they might have taken on aspects of real animals. Think of it this way: You know the story of little red riding hood, right? The wolf meets the little girl on her way to Grandma's house, then runs ahead and swallows the grandma whole, then tricks the little girl into coming close enough to swallow too. That story was never intended to be about a real, actual talking wolf but a warning to children not to talk to strangers. (There are plenty of other things going on in that story, but that's the main takeaway.)

In other words, it's quite likely that the sirrush was never meant to be anything but a creature of mythology, a glorious pet for a god. It's also possible that it was based on a known creature, just as the talking wolf in Little Red Riding Hood is based on the real wolf that can't talk.

If that's the case, what might that animal be?

People have made many suggestions over the years. The zoologist Willy Ley even suggested it was a modern dinosaur, possibly the mokele-mbembe. That was before the mokele-mbembe stories were widely recognized as hoaxes. Other people have suggested it was an animal called a Silesaurus.

Silesaurus grew up to 7.5 feet long, or 2.3 meters, and does kind of resemble the Ishtar Gate sirrush. It was slender with a long tail, long neck, and long legs. It had big eyes and probably mostly ate insects and other arthropods.

Silesaurus had traits found in dinosaurs but it wasn't actually a dinosaur, although it belonged to a group of animals that were ancestral to dinosaurs. But it probably had one trait that puts it right out of the running to be the model for the sirrush, and that is that paleontologists think it had a beak. This wouldn't have looked like a bird's beak but more like a turtle's, but it would have made the shape of the head very different from the snake-like head of the sirrush. Silesaurus probably pecked like a bird to grab insects. It also had stronger rear legs than front legs, as opposed to the sirrush that was depicted with birdlike rear legs but muscular lionlike front legs.

Silesaurus also lived 230 million years ago in what is now Poland in Europe, so there's just simply no way that it survived to modern times, no matter how much it superficially resembles the sirrush.

Ley also claims that the sirrush was the same dragon mentioned in the Bible, in a story called "Bel and the Dragon" in the extended Book of Daniel. Daniel slays the dragon by feeding it cakes made from hair and pitch. But there's actually no connection between the sirrush and the dragon in this story.

One very specific detail of the sirrush is its forked tongue. This is a snake-like trait, of course, but some lizards also have forked tongues. Could the sirrush of mythology be based on a large lizard? For instance, a type of monitor lizard?

The largest monitor lizard species is the Komodo dragon, which can grow some 10 feet long, or more than 3 meters, but there are smaller, more common species that live throughout much of Australia, Africa, and southern and southeastern Asia. That includes the Middle East.

The desert monitor was once fairly common throughout the Middle East, although it's threatened now due to habitat loss. It can grow up to 5 feet long, or 1.5 meters, and varies in color from light brown or grey to yellowish. Some have stripes or spots. It eats pretty much anything it can catch, and like many monitor species it's a good swimmer. It hibernates in a burrow during the winter and also spends the hottest part of the day in its burrow. Like other monitor lizards it has a forked tongue and a flattish head. And it has a long tail, fairly long, strong legs, and a long neck.

If the sirrush was based on a real animal, it's a good bet the animal was the desert monitor. That doesn't mean anyone thought the sirrush was a desert monitor or that we can point to the desert monitor and say, "Ah yes, the fabled sirrush, also called Mušḫuššu." But people in Mesopotamia would have been familiar with this lizard, so a larger and more exaggerated version of it might have inspired artists and storytellers.

The Ishtar Gate has been partially reconstructed from bricks found in archaeological digs. It's absolutely gorgeous. Also, the desert monitor is totally adorable.

∾

Milton Lizard

THE MILTON LIZARD is a mystery from Kentucky in the United States. In July of 1975, Clarence Cable, co-manager of the Bluegrass Body Shop in Milton, Kentucky, saw a huge lizard behind some junked cars. His brother saw the lizard a few days later. Then Cable saw it again the next day. He threw a rock at it and the lizard vanished into some brush, and that was the last anyone ever saw of it.

Cable said the lizard may have been as much as 15 feet long, or 4.6 meters, with black and white stripes overlaid with speckles, and looked like a monitor lizard. That's probably what it was, too. Monitor lizards are often kept as pets. If one escaped or was turned loose in the area, it would have survived until the weather turned too cold.

If you have a pet monitor lizard, keep it safe and warm and don't let it wander around in strange junkyards.

LAMBTON WORM AND THE LINDORM

Until the early 13th century or so, the word dragon wasn't part of the English language. Instead, the word worm was used to mean any animal that was snaky in shape. Old stories of dragons in English folklore are frequently snakier than modern dragons. For instance, the Lambton worm.

The story goes that a man called John Lambton went fishing one Easter Sunday instead of going to church, and as punishment he caught not fish but a black leech-like creature with nine holes on each side of its head. He flung it into a well in disgust and joined the crusades to atone for fishing on the Sabbath.

While he was gone, the worm grew enormous. It killed people and live-stock, uprooted trees, and even blighted crops with its poisonous breath. It couldn't be killed, either, because if it was chopped in two, its pieces rejoined.

When John Lambton returned from the crusades seven years later and found out what had happened, he sought the advice of a local wise woman, who told him what to do. He covered a suit of armor with sharp spines, and wearing it, lured the worm into the river Wear, where it tried to squeeze him to death. But the spines cut it up into pieces that were swept away by the river so they couldn't rejoin. The end.

The sea lamprey has seven little holes behind each eye called branchial openings. It's also eel-like and can be partially black, and can grow up to 3 feet long, or about 91 centimeters. Sea lampreys can live in freshwater and would have been common in the Wear until the industrial revolution of the late 19th century polluted the river. That's not to say that the Lambton worm was anything but a creature of folklore, but the sea lamprey may have inspired the story.

While the sea lamprey, crocodiles, and big snakes undoubtedly influenced dragon lore, something else did too. There's a reason dragons are so often supposed to live in caves, for instance. Caves are good places to find fossils of huge extinct animals.

In Klagenfurt in Austria there's a monument of a dragon, called the lindorm or lindwurm, that was erected in 1593 to commemorate the finding of a dragon's skull.

The story goes that a dragon lived near the lake and on foggy days would leap out of the fog and attack people. Sometimes people could hear its roaring over the noise of the river. Finally the duke had a tower built and filled it with brave knights. They fastened a barbed chain to a collar on a bull, and when the dragon came and swallowed the bull, the chain caught in its throat and tethered it to the tower. The knights came out and killed the dragon.

The original story probably dates to around the 12th century, but it was given new life in 1335 when a skull was found in a local gravel pit, which everyone assumed was the dragon's skull. The monument's artist based the shape of the dragon's head on the skull.

The skull is still on display in a local museum, and in 1935 it was identified as that of a wooly rhinoceros.

The statue

Other dragon stories probably started when someone saw huge fossils they couldn't identify. Dragons, after all, can look like just about anything. Stories of benevolent dragons living on Mount Pilatus in Switzerland may have been inspired by the many pterodactyl fossils found in the area.

In 1421 a farmer saw a dragon flying to Mount Pilatus, so close to him that he fainted. When he woke, he found a stone left for him by the dragon, which he claimed had healing properties. The dragonstone is in a local museum these days and has been identified as a meteorite.

LIGHTBULB LIZARD

The story of this little lizard starts in 1937, when a biologist named Ivan Sanderson was collecting freshwater crabs on a mountaintop in Trinidad. They were probably mountain crabs, also called the manicou crab, which is actually a pretty astonishing animal on its own. It's a freshwater crab that doesn't need to migrate to the ocean to release its eggs into the water. Instead, the female carries her eggs in a pouch in her abdomen. The eggs hatch into miniature crabs instead of larvae, and they stay in her pouch until they're old enough to strike out on their own.

The mountains of Trinidad are made of limestone, which means they're full of caves, and Sanderson was reportedly catching crabs in an underground pool or stream. He noticed a flash of light in the darkness and naturally went to find what had made it. All he found was a little lizard hiding under a ledge. It looked kind of like a brown skink and was pretty boring, but when the lizard turned its head, Sanderson saw a flash of dotted lights down both its sides. When he caught the lizard and examined it while it was sitting in his hand, it flashed its lights again.

Sanderson knew he'd found something extraordinary, because lizards don't bioluminesce. Lots and lots of marine animals do, and some terrestrial invertebrates like lightning bugs and glow-worms, but no terrestrial vertebrates.

Sanderson took the lizard back to his camp, where he and his team observed it in different situations to see if it would light up again. They moved it to warmer areas and colder ones, made loud noises nearby, even tickled it, and they did indeed see it light up a few times. The light came from a row of tiny eyespots along its sides, from its neck to its hips. It had one row of these spots on each side and each spot looked like a tiny white bead. The greenish-yellow flashes of light seemed to shine through the spots, as Sanderson said, like "the portals on a ship."

Sanderson sent the lizard to the British Museum in London where another zoologist studied it and discovered that it was actually a known species, but apparently very rare. Only two specimens had ever been caught, one a juvenile and one an adult female. The lizard Sanderson caught was male, and it turns out that only adult males have these little eyespots. Sanderson later caught seven more of the lizards.

Shreve's lightbulb lizard grows around 5 inches long at most, or 13 centimeters, not counting its long tail. It has short legs, a pointy nose, and broad, flat scales on its back and sides. It's mostly brown in color. It lives in high elevations in the Caribbean island of Trinidad and Tobago, which is just off the coast of Venezuela in South America. Unlike most reptiles, it prefers cool climates. While it turns out that it's not actually very rare, it's also hard to study because it lives in such remote areas, so we don't know much about it. It may be nocturnal and it may be semi-aquatic. It certainly lives along mountain streams, where it eats insects and other small animals.

Ivan Sanderson made a lot of sketchy cryptozoological claims later in life, but he was a biologist with a good reputation as a field scientist. More importantly, he wasn't the only one who saw the lizard light up. The problem was that after the initial reports of Sanderson and his team, no one else saw the lizard flash light.

The British Museum zoologist, H.W. Parker, who studied the first lizard Sanderson found, was actually the scientist who had originally discovered the lizard a few years before. He was very interested in the little portholes along the male lizard's sides and studied them carefully. But he couldn't find anything about them that indicated how they lit up. Each tiny eyespot consisted of a transparent center spot with a ring of black skin around it. The eyespots didn't contain glowing bacteria, specialized nerve endings, ducts, reflecting structures, or anything else that he could think of that might cause a flash of light.

Other zoologists examined the lightbulb lizard over the next few decades and none of them saw it emit light either. By 1960 no one believed it was bioluminescent.

Fortunately, there've been some more recent studies. The lizard has been reclassified several times and its current name is *Oreosaurus shrevei*. While I would like to think the name comes from the white-appearing center of the eyespot with black pigment around it like an Oreo cookie, the name Oreosaurus is older than the cookie and as far as I can tell it means mountain lizard.

Experiments conducted in the early 2000s finally figured out just what is going on with the lightbulb lizard. Sanderson was right: he and his colleagues really did see light coming from the eyespots. But it's reflected light, not light emitted by the lizard itself. The white dots in the middle of the eyespots are reflective at some angles. Not only that, but when the lizard feels threatened, the skin around the white dots becomes even darker, which makes the reflection seem brighter. It's partly optical illusion, partly just optics.

The big question now is why the lightbulb lizard has these reflective spots at all. The female doesn't have them. That suggests that the male uses them in some way to attract a mate, but only the lightbulb lizard knows for sure.

VIETNAMESE MYSTERY SNAKE

I n 1968, during the Vietnam War, someone in the United States Naval Medical Research Unit discovered a small snake in central Vietnam. It was unusual enough that they decided to save it for snake experts to look at later, but things don't always go to plan during wartime. The specimen disappeared somewhere along the line. Fortunately, there were photographs.

The photos eventually made their way to some biologists, and in 1994 a paper describing the snake as a new species was published by Wallach and Jones. They based their description on the photos, which were good enough that they could determine details like the number of scales on the head and jaw. They named it *Cryptophidion annamense* and suggested it was a burrowing snake based on its characteristics.

Other biologists thought Cryptophidion wasn't a new species of snake at all. In 1996 a pair of scientists published a paper arguing that it was just a sunbeam snake. The sunbeam snake is native to Southeast Asia, including Vietnam, and can grow over 4 feet long, or 1.3 meters. It's chocolate-brown or purplish-brown but has iridescent scales that give it a rainbow sheen in sunshine. It's a constricting snake, meaning it squeezes the breath out of its prey to kill it, but it only eats small animals like frogs, mice, and other snakes. It's nocturnal and spends a lot of its time burrowing in mud to find food.

Wallach and Jones, along with other scientists, argued that there were too many differences between the sunbeam snake and Cryptophidion for them to be the same species. But without a physical specimen to examine, no one can say for sure if the snake is new to science or not. If you live in or near Vietnam and find snakes interesting, you might be the one to solve this mystery.

~

Tsuchinoko

THE TSUCHINOKO of Japan is supposed to be a short but wide-bodied snake with horns above its eyes, a broad head with sensory pits, and a thinner neck. Its pronounced dorsal ridge makes it seem somewhat triangular in shape instead of rounded like most snakes. Think Toblerone bar, but snake. It's also said to be able to jump long distances.

Some cryptozoologists suggest it might either be an unknown species of pit viper or a rare mutant individual of a known pit viper species. Stories of tsuchinoko sightings go back centuries, although more recent accounts describe it as a more ordinary-looking snake with a big bulge in its middle as though it has just swallowed something that it hasn't digested yet.

In 2017, a Tumblr post inspired a meme about the tsuchinoko. It's a picture of three cats staring at a fat lizard with the legs Photoshopped out and the caption "tsuchinoko real," which I'm sure you can agree is meme *gold.*

PART TEN
DEMONS AND SPECTERS

This section covers mysteries that skirt along the edge of reality. What made the so-called devil's footprints in England? Was the Dover demon of North America a creature of earth or something more sinister? It's Halloween every day in this section and we're on a monster hunt.

DEVIL'S FOOTPRINTS

The winter of 1855 was especially bitter in England. Around Devon, the rivers froze solid and temperatures stayed below freezing almost every day and night from January to March. On the night of February 8 it snowed, but towards dawn a brief thaw turned the falling snow to rain before the temperature dropped again and a frost fell. When residents of Devon woke on the morning of February 9, they found some 4 inches of snow on the ground, or 10 centimeters. They also found small hoofprints everywhere.

These weren't ordinary hoofprints. A donkey or pony hadn't gotten loose during the night and wandered around. Some of the prints did look like a donkey's, but some appeared cloven, more like a large goat's hoof. And the stride was short, only about 8 inches between most prints, or a little over 20 centimeters, sometimes about double that. Besides, the prints appeared in places where a donkey couldn't possibly have left prints: on rooftops, inside gardens with tall walls and locked gates. Even a nimble goat couldn't have managed that without someone hearing a goat bounding around. Sometimes a line of prints would walk right up to an obstacle, like a haystack or hedge, and continue on the other side as though the obstacle didn't exist. Tracks began or ended abruptly as though the animal had dropped from or flown into the sky.

There were untold thousands of the prints. Some villages had prints in almost every yard. They appeared in churchyards among gravestones, in gardens and on doorsteps, in fields and roads. They meandered from place to place or sometimes continued in a straight line. And they appeared to be made not by a four-footed animal but by something walking on its hind legs, placing one hoof nearly in front of the other.

People tracked some of the prints for miles without coming across any clue as to what had made them. A few forward thinkers made sketches of the prints and jotted down notes. By February 13, reports of the strange footprints had made it into the local newspapers.

Beyond the often maddeningly vague newspaper accounts, most of what we know about the hoofprints comes from the Reverend H.T. Ellacombe, who was vicar of the parish of Clyst St George from 1850 to 1885. He collected letters and sketches and made his own notes about the event, since some of the prints appeared in his own rectory grounds. Local historian Major Antony Gibbs discovered Ellacombe's bundle of notes and letters in 1952, tucked away in a church office gathering dust.

But a series of letters published in 1855 by the Illustrated London News has been more influential than Ellacombe's information. The letters were written by someone who signed himself "South Devon," and we know from Ellacombe that South Devon was a 19-year-old local man whom Ellacombe called "young D'Urban."

William D'Urban's letters were exciting, to say the least. If you've heard anything about the devil's footprints before, it was probably mostly details from D'Urban's account. According to him, all the prints were identical in size, the stride likewise did not vary, and the prints were one unbroken trail at least 40 miles and as much as 100 miles in length, or 64 to 160 kilometers. This has sometimes been garbled in later retellings as a perfectly straight trail 100 miles long. D'Urban was also the one who claimed the prints continued from one side of the River Exe to the other side, 2 miles distant, or 3.2 kilometers. It's not clear if the river was frozen at this point, although it was frozen so solid by late February that an enterprising stove manufacturer ran pipes from the gas main onto the river ice and cooked an entire dinner for 30 on it while people skated all around him and probably tripped over the gas pipes. Moreover, the river is an estuary of the sea so has tides, and at low tide it's barely a few hundred yards wide in some areas, or say 200 meters, and barely 4 feet deep, or about 1.2 meters.

Even at the time, D'Urban's account was refuted by other locals, whose letters responding to South Devon's letters were printed in follow-up issues of the paper. Newspapers back then were like really slow social media. People wrote letters in response to other letters they'd seen in the newspaper, and other people wrote letters in response to those letters. Old timey people really needed Facebook—and cameras, because we don't have very many sketches of the footprints and the ones we do have aren't very detailed.

So what did the tracks really look like? As far as we know, most of the tracks were about 4 inches long, or 10 centimeters, and 2.75 inches across, or 7 centimeters. They did vary in size and shape from place to place, which argues that more than one animal made them and that hoaxers weren't involved, since hoaxers would leave identical prints. When you hear the word hoofprint it's easy to think of a crisp, well-marked round hoof, maybe even with a horseshoe, but these prints were kind of wobbly in shape—not unexpected since they were all somewhat distorted by the night's thaw and refreeze.

One of the people who wrote in to denounce some of D'Urban's details was a Reverend G.M. Musgrave, vicar of Exmouth, and one of the things Musgrave also mentions is that he himself had suggested to his parishioners that the tracks were made by kangaroos escaped from a private menagerie. But, he admits, he didn't actually believe this, he was only trying to stop his parishioners from believing that the devil had walked through their town.

The devil only started getting blamed for the footprints once it was clear no one really knew what had caused them. Lots of animals were suggested as culprits, most of which were about as likely as Musgrave's kangaroos. Among the suggestions were badgers, rats or mice, hares, wolves, cats, monkeys, toads, or various birds. One anonymous letter-writer said that a friend had examined the tracks, noted that some of them showed claw marks, and suggested the animal might be an otter—mostly as a way to explain how the trail passed under low branches without disturbing them and through a 6-inch, or 15-centimeter, pipe.

Other suggestions were even more outlandish, like the runaway balloon trailing a rope theory. Or the complex and largely irrational theory proposed in 1973 that seven Romany tribes conspired to lay the tracks in one night using stilts made from stepladders, in an attempt to scare some other tribes away. Or the 1972 theory that UFOs were measuring...something...with lasers and the tracks were left as a result, by lasers. Measuring things.

Leaving aside the theories that are clearly farfetched, like animals escaped from menageries and UFOs, and going with the assumption that whatever left the tracks was likely a real animal native to England, what might have left the devil's footprints? I'm going out on a limb and suggesting maybe it wasn't the devil.

Badgers, otters, and wolves leave tracks much too large to fit the descriptions. Toads are cold-blooded and wouldn't be active in the snow. Birds don't leave miles of prints in snow at night, not even owls hunting mice on the ground, as they sometimes do. The tracks of deer would probably be recognized no matter how distorted the melting snow might have made them, and there are no reports of dew claw marks that deer prints show.

What about cats? Cats leave small neat footprints in snow with prints nearly in front of each other. With the brief thaw, feral cats might be out hunting for mice and other animals around houses and gardens, exactly where many prints were found. Cats can climb well, and a small cat might be able to accomplish some of the astonishing feats reported, like getting through dense hedges or larger pipes. And we do have a witness whose report is interesting. A tenant of Aller Farm in Dawlish, the only person we know to have been outside during the night in question, said that his cat had left tracks in the snow and that the thaw and rain melted them, after which they froze again into small hoof-like shapes. It's possible that at least some of the prints were made by cats.

Rats sometimes hop through snow on all four feet, leaving deeper impressions that do look remarkably like the hoofprints seen. Rats can also get through quite small spaces and climb well. The main drawbacks of this theory are that hopping rats leave clear tail prints and rats don't hop for miles. Rats also usually leave prints larger than the ones found. But again, it's possible that at least some of the prints were made by rats.

Finally, what about mice? When I was a kid, this argument seemed ridiculously weak. I had pet mice. I knew there was no way a mouse could leave a horseshoe shaped print in the snow. But I was only familiar with pet white mice and house mice. There's a type of mouse common throughout Europe that I think might be our culprit. Let's find out why, and learn about the wood mouse.

The wood mouse, also called the long-tailed field mouse, is a cute little rodent with a long tail, sandy-brown or orangey fur, white or gray belly and legs, and big ears. Not counting its tail, it's about 2.5 to 6 inches long, or 6 to 15 centimeters, and its tail can be as long as its body. It mostly eats seeds and nuts, although it will also eat roots, shoots, berries and other fruit, moss, fungi, snails, and insects when seeds aren't available.

Like many rodents, it discovered a long time ago that humans are useful nuisances, so it frequently lives around houses and barns, although not usually *in* houses. It generally lives in burrows it digs in fields, gardens, or among the roots of trees, although sometimes it will make its nest in bird-houses, hollow logs, or in thick vegetation. The nesting chamber of a mouse's burrow is lined with leaves, grass, and moss, and it may also dig chambers where it stores extra food.

In warm weather wood mice aren't very social, but in winter they will sleep in pairs or groups to stay warm. They don't hibernate, but in especially cold weather they become torpid. They're nocturnal animals, good climbers, jumpers, and swimmers.

While it forages, a wood mouse will pick up small items like leaves and twigs and place them in conspicuous locations to mark certain areas. As far as researchers know, wood mice and humans are the only animals to mark trails with items, known as way-marking. A mouse's typical winter territory is around half an acre in size, or 2,000 square meters.

All this is interesting, but why do I think the devil's footprints were mostly made by wood mice? Well, wood mice often travel by hopping on all four legs. They're built like tiny kangaroos, with long hind legs and comparatively short forelegs. Unlike a rat, a jumping wood mouse doesn't leave much of a tail mark in snow. It can also keep up this hopping gait for a long time, which it would do since it's a more efficient way to travel through snow taller than the animal is high. It jumps with its feet together so the print it leaves behind roughly resembles a V shape where the two sides of the V don't connect. Any amount of thawing and refreezing can turn that print into a cloven hoofprint or a donkey-like hoofprint.

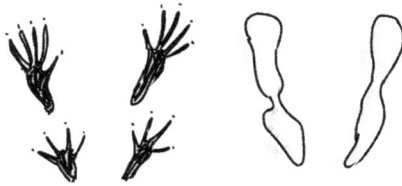

Wood mouse prints before and after thaw

Moreover, mice can get through extremely small holes and pipes, can burrow straight through haystacks, can hop across roofs without making noise. Where people reported finding prints that vanish in the middle of open fields, the mouse could have disappeared into a burrow, been picked off by an owl, or just stopped hopping and started walking, leaving footprints so small and shallow they likely didn't survive the thaw.

But why were there so many prints on this particular night? Remember, the winter had been harsh but that particular night there was a brief thaw. It's probable that even slightly warmer weather would bring hungry mice out in droves to forage. The unusual weather conditions that distorted otherwise barely noticeable tracks into hoofprints, and human nature, did the rest.

But if that's the case, why haven't people reported seeing the same mysterious prints at other times? Actually, they have, both before and after 1855.

The earliest account anyone has found in the newspapers was an 1840 report in the London *Times* of strange prints in Scotland. Other accounts date from the 1850s, 1890, the 1920s, the 1950s, and so on until 2009.

Some of these accounts are of much larger prints, some don't match up with the hoofprints seen in 1855, but some sound similar. In 1957, for instance, when Lynda Hanson in Hull was a child, a line of cloven hoofprints 4 inches long, or 10 centimeters, and 12 inches apart, or 30 centimeters, appeared in her family's garden in snow that had fallen overnight. They vanished in the middle of the garden. Ms. Hanson noted that the family dog didn't bark.

Another interesting report comes from a sighting in late 1962 or early 1963. Zoologist Alfred Leutscher, writing in the April 20, 1965 edition of *Animals* and expanding on a talk he gave to the Zoological Society of London about the sighting, explains some tracks he found in Epping Forest. I'll quote from his description. "It was during a search for snow tracks in Epping Forest, in the severe winter of 1962-3, that I came across dozens of trails of the wood mouse, each consisting of small 'V-shaped' marks regularly spaced out and conforming to the measurements which were given a hundred years ago. When I found them I was totally unaware of their significance."[1]

There are problems with this, of course. While the account says the tracks were identical to those reported in 1855, they're described as V-shaped rather than hooflike. I have no doubt Leutscher's prints were from wood mice, but whether they were the same type of thing seen in 1855 in Devon, we can't know for sure since the reports from the 1855 sighting are so unclear. The 1855 thaw and refreeze might be the cause of the differences, however.

In 2009, Jill Wade of North Devon woke up to snow and found a line of hoof-like prints across her garden. A zoologist who examined the prints suggested they might be those of a rabbit or hare, although since the prints were only 5 inches long, or 12.5 centimeters, that would have to be a little baby bunny. But the great thing in this case is we have photographs. It definitely looks like a hoofprint—and it also looks like little animal legs made it.

One interesting thing. The wide part of a wood mouse's print, the one that would make the rear of a hoofprint, is actually at the animal's front. So anyone following the devil's tracks in 1855 was following them backwards.

AHOOL

Bats are grouped into two basic types, microbats and megabats. Microbats are usually small, flat-faced with big ears, and use echolocation to catch insects at night. Megabats are typically larger, with limited echolocation abilities and longer muzzles, and they often eat fruit. The biggest bat alive today is probably the great flying fox, which lives in New Guinea and nearby areas. Its wingspan can be nearly 6 feet across, or 1.8 meters. The golden-crowned flying fox, which lives in the Philippines, is very nearly as large, with a wingspan of over 5.5 feet, or 1.7 meters. Both are megabats.

But there's a story of a bat much larger than these flying foxes, called the ahool.

The first official report of an animal called the ahool comes from western Java in 1927. Naturalist Ernst Bartels was in bed but still awake when he heard a loud call that sounded something like "a-hool!" Bartels rushed outside with a flashlight in hopes of seeing what animal had made the call. He heard it again farther away, then again almost out of earshot.

As it happens, Bartels had grown up in western Java and knew about the legend of the ahool. It was supposed to be a monstrous bat, its wingspan some 12 feet across, or 3.6 meters. Its face was monkey-like with large eyes and it was supposed to have feet that pointed backwards. During the day it lived in caves hidden behind waterfalls but at night it flew out and scooped fish from the river.

Bartels did more research into the ahool legend and eventually Ivan Sanderson, a cryptozoologist who had his hand in everything back in the day, contacted him with his own account. In 1932 Sanderson said he had seen a gigantic black bat in western Africa one night. He and his companion, naturalist Gerald Russell, had been searching for tortoises in a river when the bat flew over them. They both estimated its wingspan as 12 feet and Sanderson said he could even see the sharp teeth in its open mouth.

I'm skeptical of Sanderson's sighting because Sanderson was *always* having remarkable cryptozoological sightings with no proof but his own say-so. No one's that lucky and unlucky at the same time. He should have carried a camera with him at all times. Bartels's story was documented by Sanderson and is therefore suspect too.

But the ahool is an interesting cryptid because its description sounds so plausible. Even the backwards feet make sense because bat feet have evolved to allow bats to hang upside down easily, which means their feet do appear to be backwards when they're right side up.

There are some discrepancies, though. Megabats all have long muzzles compared to microbats. The ahool is specifically described as having a flat face like a human or a monkey. While megabats aren't all completely frugivorous—that is, they don't all eat fruit—none of them are known to eat fish. Some microbats do specialize in catching fish, though, and those bats all have longer snouts than typical insectivorous bats, although not as long as megabat snouts. The ahool is also said to stand or sit upright on the ground occasionally. While microbats sometimes do stand upright, megabats never do.

Sanderson proposed that the ahool may actually be an enormous microbat. Some microbats are actually pretty big, including the carnivorous ghost bat, also called the false vampire bat, which lives in parts of northern Australia. Its wingspan is almost 20 inches, or 50 centimeters. This is much larger than the smallest megabat, the spotted-winged fruit bat, which has a wingspan of only 11 inches, or 28 centimeters.

But microbats don't make a lot of noise. It's megabats who honk and call to each other rather than just squeaking. The ahool is supposedly named for its loud cry, the one Bartels heard. Remember that Bartels never saw the animal he heard.

Let's look at the possibilities logically.

Possibility one: the ahool is a real animal, exactly or mostly as described, with aspects of both megabats and microbats. It's either incredibly rare or extinct these days, which would explain why there aren't more sightings. It would have to be an animal completely new to science.

Possibility two: the ahool is a real animal but it's not well known because it's so seldom seen. It only seems to be a mixture of microbat and megabat because people who saw it made assumptions of its appearance based on what they know about bats. Some of those details are from micro-bats and some from megabats, and the actual animal may not look anything like its description in folklore. In this case, it's probably a megabat and may be one already known to science.

Possibility three: the ahool is an animal entirely of folklore and myth, described as similar to various bats familiar to locals but enormously large. If the ahool's call is actually that of a rare or migratory bird, seldom heard and therefore mysterious, it might have inspired the myth.

I can't even make a guess as to which possibility might be the most likely, and that's pretty awesome. Hopefully researchers from Java will find out more about the ahool soon.

DOVER DEMON

D over, Massachusetts is a small town in northeastern North America, only about 15 miles, or 24 kilometers, from the city of Boston. Currently just over 6,000 people live there, up from about 5,000 people in 1977. It's an affluent town with good schools, a small museum, and a number of historic homes. But it's most well known for something weird that happened over forty years ago.

On the night of April 21, 1977, three 17-year-old boys were driving along Farm Street on the outskirts of town. Bill Bartlett was driving with his friends Mike and Andy in the car with him. This was long before the internet or video games or even cable TV were invented, so they were just driving around and talking. It was a Thursday night, cooling down after an unusually warm day for late April in that part of North America.

Around 10:30 pm, as they passed along a low stone wall on the left side of the road, Bill noticed an animal of some kind climbing on the stones. He thought it was a dog or even a cat at first, but then the car's headlights lit it up. Bill saw the creature turn its head and stare into the light.

The creature was definitely not a dog. It had big round eyes that shone like orange marbles, as Bill described it later. He estimated it would have been 4 feet tall if it had been standing upright, or 1.2 meters. It had peach-colored skin that looked like it might have a rough texture, which Bill later described as looking like wet sandpaper or a shark's skin. Its body was thin, its arms and legs were very long and thin, and it had a thin neck. But its head was oversized and oddly shaped. Bill described it as shaped like a melon, but to me it always sounds more like Snoopy the dog's head. Snoopy has a round head and a big oblong nose in comparison to a little body. But this creature wasn't a cartoon character, it was real. Bill said it had long, thin fingers and toes that it wrapped around the rocks as it climbed over them. But while it did have those big glowing eyes, it didn't appear to have ears, nose, or mouth.

The other two boys in the car were talking and didn't notice the creature as the car passed it going somewhere around 45 miles per hour, or 72 kilometers per hour. Bill was naturally freaked out and after about three quarters of a mile, or a little over a kilometer, he stopped the car to tell his friends what he'd seen. They talked it over for a good 15 minutes before deciding to turn around and go back to look for the creature. They didn't see it again so Bill dropped his friends off at their homes, then went home himself.

Bill's father noticed he seemed upset and Bill admitted he'd seen something that had spooked him. He made a drawing of the creature and later made another drawing and a watercolor of it. Bill was a good artist and in fact when he grew up he became a professional artist.

A few hours later that same night, at about 12:30 am, another boy, -year-old John Baxter, was walking home from his girlfriend's house on Miller Hill Road. Miller Hill Road intersects Farm Street, the road where Bill saw the creature, and John was about a quarter mile, or almost half a kilometer, away from where the two roads met when he noticed another person on the road ahead. He noticed the figure's large head and thought it was a friend who lived nearby, a boy who actually had a deformed head due to a childhood illness. John called out to him but didn't get a response. As John came closer, he noticed how small the other figure's body was in comparison to its head and realized it wasn't his friend. He thought it might be a young child.

The figure suddenly ran off the road. John heard it run down a small embankment at the edge of the road and into the trees. He chased it but stopped at the bottom of the embankment, since he hadn't realized there was a creek at the bottom and almost fell in. The creature must have jumped the creek because John specifically mentioned that he didn't hear it splash through the water.

At this point John got a good look at the creature, since it had stopped about 30 feet away, or 9 meters, and was looking back at him. It was in silhouette against an open field, with its head about level with John's because of the way the ground sloped. It had a small, slender body, long, thin arms and legs with long, thin fingers and toes, and an oversized head with the same unusual shape that Bill reported. John said that the creature was standing on a rock and he could see that its long toes were wrapped around the rock, while it had also wrapped its long fingers around the trunk of a small tree.

Naturally, John got spooked at that point and backed off. He hurried to Farm Street and found someone to give him a ride home instead of walking the rest of the way. Later he made a sketch of what he'd seen.

The next night, a 15-year-old girl named Abby Brabham reported seeing something weird on Springdale Avenue. Springdale also joins Farm Street and is less than a mile from Miller Hill Road. Abby was riding in the car with an 18-year-old friend, Will Taintor, who was driving her home, when she noticed something crouched next to the road at the edge of a bridge. The road doesn't have any structure that could be called a bridge today, but there are a lot of swampy areas nearby so there may have been a low bridge there in 1977. At first Abby thought she was looking at an ape, but it had a tan body without hair, a large watermelon-shaped head, and glowing green eyes. Will saw the creature too, although not as good a look as Abby. Abby estimated that the creature was the size of a goat.

That's that. The creature was never seen again.

OH MY GOSH THAT IS SO CREEPY.

In a case like this, where we're presented with accounts from four people who saw something truly weird and report specific details that tally with each other, we can look at it logically this way. It's either a hoax, a known animal that wasn't identified at the time, an unknown animal, or something supernatural.

Let's start with the assumption that it was a real animal, either known or unknown. What kind of animal might it be?

People have suggested that the animal might have been a moose calf that got separated from its mother and was blundering around scaring teenagers. But a moose as the culprit doesn't make any sense. Moose have big ears and nostrils, hooves instead of long fingers and toes, and a calf's head actually appears small in relation to its bulky body and extremely long legs. But most importantly, there weren't any moose living anywhere in Massachusetts in 1977. Now that there are moose in Massachusetts, you'd think people would start seeing the Dover demon again if it was actually a young moose, but that hasn't happened.

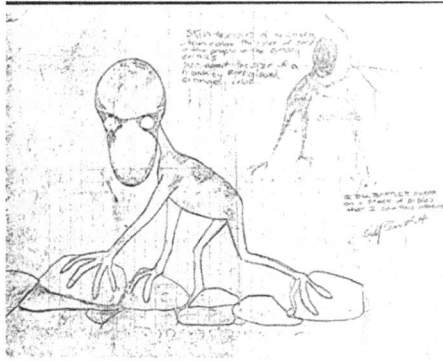

Bill Bartlett's original drawing

A more likely possibility is a young black bear. Bears do stand and some-times walk on their hind legs, especially young ones. But bear paws are broad and flat and their toes are close together, sort of like a person's toes. Not only that, bears have large ears. And, of course, they have thick black fur. Bears do get a type of mange that can cause them to lose their hair, and a young bear with a case of mange that advanced would probably also be sick and possibly very thin. But hairless bears, in addition to looking very sad, appear to have small heads, and their ears stick out even more than usual since they're not half hidden with fur.

Could the creature be a primate of some kind? Abby said she thought she was looking at an ape at first, while in his interview John said he thought the creature might be a monkey. Obviously neither monkeys nor apes would ordinarily be found in Massachusetts, but exotic animal laws were more lax in 1977 than they are now. Not only that, there are some primate research facilities in and near Boston.

A monkey would have the long, thin limbs and slender body seen by witnesses, and it could wrap its fingers and toes around rocks. But there are problems with the monkey hypothesis. Monkeys generally have small heads, not oversized ones. Most have tails. Monkeys are also diurnal so wouldn't be running around at night, and even if it was out at night for some reason, it would likely stay in the trees or climb a tree if someone frightened it.

Apes, of course, have no tails and they have relatively long limbs and fingers and toes. But it's much less likely that an ape would escape from someone's home or from a research facility and not be reported missing. Apes are expensive and can be dangerous. The research facilities in the area don't keep apes, just monkeys of various kinds.

It couldn't have been a gorilla since a gorilla's face is always gray or black with pronounced nostrils, and gorillas are much larger than what witnesses reported seeing, with a bulky body. Chimps are closer to the right size, but chimps don't have big heads compared to their bodies—in fact, they're proportioned more like humans but with smaller heads.

But what about an orangutan? Leaving aside the issue of where it came from and why it was never reported missing, could an orangutan have been the Dover demon? Dominant male orangutans develop large cheek pads and throat pouches that can make their heads look quite large, and most orangs have orange fur that might look like tan textured skin in the dark. But orangutans have noticeable mouths and nostrils, plus their eyes are much closer together than the drawings indicate.

There's something else that indicates the Dover demon probably wasn't a primate. Many animals have a reflective layer in the eye that causes eyeshine at night. It's called the tapetum lucidum and it helps animals see better at night. But humans, apes, and monkeys are diurnal animals and don't have that particular night-time adaptation. All the witnesses said that the creature's eyes glowed, presumably with reflected light.

That brings us to Abby's sighting. Abby reported that the creature she saw had glowing green eyes, whereas Bill and John both said the creature they saw had orange glowing eyes. Abby refused to change her story, either, and was adamant that the eyes she saw were green.

Different animals have different-colored eyeshine, and individuals of the same type of animal may have different color eyeshine depending on lots of factors. But the color of an individual animal's eyeshine doesn't change from one night to the next. Is it possible that Abby didn't see the same creature that Bill and John did? She didn't get a good look at it, and

Will barely got a glance. I've always wondered why Abby said the creature she saw was the size of a goat. While that is a good description, unless Abby was familiar with goats and saw them frequently, it's surprising that she didn't describe it as the size of a big dog. I wonder if Abby actually saw a tan or light brown goat by the side of the road but didn't recognize it consciously. Goats do have green eyeshine.

Then again, Abby described the creature's head as very big and very weird, specifically watermelon-shaped, and she specified that it had a tan, hairless body and round eyes. All these things tally with what the others reported.

So if it wasn't a known animal, could it have been an unknown animal, something new to science? Let's assume it was, for now, and try to figure out what kind of habitat an animal of the Dover demon's appearance would belong in.

However strange it appeared to the witnesses, the creature had four legs and a head. That means it fits the general body scheme of a tetrapod and a vertebrate. Witnesses reported seeing both fingers and toes, so it wasn't a bird. It appeared to have a tapetum lucidum so it must be nocturnal to at least some degree. It was able to move around on land rapidly, apparently could walk on its hind legs alone if necessary since John Baxter mistook it for a person on the road, but it also seemed more comfortable when its front legs were braced against something, such as a rock or a tree. John didn't hear it splash in the water when it ran down the embankment, so it could jump as well as run. It also didn't escape by climbing a tree or hiding in the water when it was frightened, so we can assume it was a terrestrial animal, meaning it was most comfortable on land. It was either a mammal, a reptile, or an amphibian, but I think we can discount the amphibian hypothesis since it avoided the water. It didn't appear to have fur and its skin had a textured appearance, which might suggest that it was some kind of reptile instead of a mammal. But all reptiles have tails and our creature didn't appear to have one. That means the Dover demon was most likely a mammal.

So we've narrowed it down to a nocturnal, terrestrial mammal that either doesn't have hair, or that was suffering from an advanced case of mange. Let's assume that it just doesn't have hair, or has hair that wasn't visible at night. The hair might have been short and dense, like seal fur, so that it appeared to be textured skin, or it might have been very finely haired, sort of like a human's body.

The witnesses agreed that the creature was pale in color, a sort of tan or peach. That's where it gets trickier, because that's an unusual trait in a nocturnal animal. Pale skin or hair reflects light, which makes the animal easier to see in the darkness. That's great if you're a skunk and want to call attention to yourself so other animals can avoid your amazing stink powers, not so great if you're trying to slink around unseen.

There are only a few habitats where it doesn't matter what color an animal is, and those are habitats where the animal can't be seen. Maybe the Dover demon ordinarily lives underground or in a cave.

Animals that live underground have to be able to dig, or else they're not going to get very far. That means they need large claws. They also tend to have smaller eyes, since they can't see far anyway and more dirt will get into large eyes. They also have short legs. So I think we can safely say that the Dover demon wasn't an animal that spent much or any time burrowing underground.

But maybe it was a cave-dwelling animal. It had big eyes and a tapetum lucidum to take advantage of low light and it navigated quickly over rocks, wrapping its long fingers and toes around them. That makes sense, because those fingers and toes, as well as the long limbs, suggest an animal that can climb. Maybe it climbed around in caves.

We seem to have found the most reasonable habitat for what we know about the Dover demon. It may be an animal adapted for climbing around in caves. Perhaps its oversized head acts as a resonant chamber to help it navigate with a type of echolocation when there's no light, the way many whales have bulbous foreheads called melons.

But it doesn't fit exactly, so let's go over the drawbacks of our cave-dwelling mammal hypothesis.

First, there aren't very many caves in Massachusetts. Only about 14,000 years ago, all of northern North America, including what is now Massachusetts, was covered with glacial ice many miles deep. The glaciers scraped away soil and the softer rock where caves form, and when they finally melted, they left exposed tough bedrock behind. What few natural caves there are in the area are quite small.

All that aside, the Dover demon was a large animal. Large animals don't live exclusively in caves because there's just not enough food for them. There are also no mammals known that live only in caves. Bats roost in caves but they come outside at night to hunt, and many other animals hibernate or sleep in caves but spend the rest of their time outside.

Now let's look at our other two choices: either it's a hoax or it's something supernatural.

I don't like to label mystery animals as supernatural. You're not solving a mystery by labeling it as a different mystery. Some people have suggested that the Dover demon was actually an extraterrestrial that got separated from its space ship, and I considered this hypothesis carefully. But the Dover demon acted like an animal and has the body plan of an animal from Earth. If it hadn't looked so spooky—if it had fur—no one would have thought it was especially unusual.

So, was it a hoax? The witnesses all went to the same high school and lived in the same small town, although Abby was actually from another small town adjacent to Dover. Bill and Will were good friends, but Bill and John only knew each other slightly. None of them went to the newspapers or tried to make a big deal of their sightings. There's some confusion as to when they realized they'd all seen the same thing and compared stories, but it seems to have happened either the next day when Bill and Will gave John a ride, on Saturday when John and Bill both attended the same party, or the following Tuesday at school.

Could it have been a thin person or a little kid with a big mask on, out at night to scare people? Of course this is possible, but why did the person quit after just three separate sightings on two nights? Reports of the sightings didn't make it to the newspapers until the following month, so it's not like there were people out hunting for the creature and the hoaxer was afraid they'd be caught.

There is one detail that concerns me about the sightings. April 18, 1977 was a new moon, which means that April 21 would have been quite dark, with only a thin crescent moon. Yet both Bill and John reported seeing the creature's long fingers and toes wrapped around rocks. Bill did see the creature in his car's headlights, but John was looking at it through the trees. Miller Hill Road doesn't appear to have streetlamps, so I don't know how John could have seen the creature's fingers and toes so clearly from 30 feet away on a nearly moonless night.

I wish we had better information about when John made his drawing. His drawing is very similar to Bill's, so much so that I wonder if John saw Bill's first and used it as a reference. While John did obviously have some artistic ability, his lines aren't as sure as Bill's and he may have been uncertain both about the details he'd seen and about his ability to draw the details accurately. He might have filled in the gaps in his memory by looking at Bill's drawing.

It does seem suspicious that three of the four witnesses were close friends: Bill and Will were buddies, and Abby was Will's girlfriend. John knew the others but wasn't friends with them, but that doesn't mean they couldn't have asked him to be part of the joke. Bill and Will were older and John would probably have been flattered to be part of the hoax.

I keep coming back to Bill's drawings of the creature. He made several, apparently illustrating what he'd seen since the poses are all similar. I'm an artist myself, and sometimes when you draw something really interesting you redo it several times because you like it and you want to improve it. You can get attached to a character you doodled randomly. I wondered for a while if Bill's drawings came first, and the story came second as a way to get more people to look at his drawings and appreciate the character he created. Or he might have been illustrating Gollum, the creature from *The Hobbit* and *The Lord of the Rings*. The books were really popular even then and the drawing does look a lot like the way Gollum is described in the books.

I finally rejected this possibility. From what we know of Bill's character, this wasn't something he would do. His friends all vouched for him being honest and apparently he was genuinely shaken by the sighting. I think he kept drawing and painting the creature because he was trying to figure out what it was.

So if it wasn't a hoax, and it wasn't a known or unknown animal, and it probably wasn't something outside of the realm of science, what's left? That's the trouble: we don't have enough information to know for sure. All we know is that four teenagers saw the creature over the course of about 24 hours, and then it was never seen again.

The more I think about it, the more I come to a single conclusion. The creature's body is short, the limbs long in proportion. There's no tail and no ears. But Bill only saw it briefly in the glare of his headlights on an extremely dark night. He saw the general shape of an animal climbing over rocks, staring into the light with its eyes reflecting the light brightly. Bears have orange eyeshine and the placement of a bear's eyes in its face matches what Bill drew.

Here's what I think happened, maybe. Bill saw a small, thin bear with an advanced case of mange. This meant it had very little hair left and its skin was probably inflamed from scratching. Mange is caused by a mite that burrows under the skin of an animal and causes intense itching. The skin and what was left of the fur would look textured and mottled. The bright eyeshine drew Bill's attention and exaggerated the size of the head in his memory, or it's possible the poor bear had a swollen face from a bee sting or another malady that made it look much larger than it really was.

John encountered the same bear later that night. It's not clear from John's description if the Dover demon was actually walking toward him or just standing in the road. John might not have been able to tell. Bears stand on their hind legs to see better, and the bear was probably alarmed at John's approach and was trying to decide what to do. When John got too close, the bear ran into the trees, then stopped to look back. That's when John saw it in silhouette, realized he was seeing something weird, and retreated.

The next evening, the creature was supposedly spotted again by Abby as Will drove her home, and it's possible that's what they saw. But by that time Bill had undoubtedly told Will about the bizarre creature he'd seen and shown him his drawing, and Will had undoubtedly told Abby. This could have influenced her brief sighting of a goat or other animal, and she thought she had seen the same thing as Bill had. She may even have seen his drawing too. Remember, people see what they expect to see. We don't do it on purpose; our brains take what we know of a situation or object or animal and fill in the gaps of what we see so we can make faster decisions about what to do.

I also think John had already seen Bill's drawing when he made his own drawing, which is why they look so much alike. I just can't believe that John could have seen the creature's toes wrapped around rocks at that distance, under the trees, in near-total darkness. I think he was influenced by Bill's drawing and added the detail without realizing he hadn't actually seen it. Bill had probably misinterpreted what he saw: namely, the bear's long claws, which might have looked like long fingers, especially if the bear had very little fur on its paws.

My theory certainly doesn't fit all the facts, primarily the presence of big bear ears which should have been quite noticeable. I did look at a lot of trail cam pictures of bears taken at night, though, and at some angles the bear's ears don't show very much or at all. I don't think my scenario is necessarily what happened, just a guess that mostly fits. We would need a lot more information to make a real identification of what creature the teenagers saw that night, and so much time has passed that it's impossible to get that information now.

Ultimately, the Dover demon remains a mystery. But even though I'm mostly satisfied that the Dover demon was a sad mangy bear, that has not stopped me from being really jumpy on my usual evening walks. Because I might be totally wrong and I'll never know.

DEVIL-PIG

The story starts in 1875, when a man named Alfred O. Walker sent a letter to the journal *Nature* about a discovery on the north coast of Papua New Guinea. It wasn't the discovery of an animal itself but a big pile of dung from an unknown animal. The dung pile was so big that the people who found it thought it must be from some kind of rhinoceros. But New Guinea doesn't have any rhinos.

The dung pile was discovered by a British expedition led by Lt. Sidney Smith and Captain Moresby from the ship H.M.S. *Basilisk*. After the report was published in *Nature*, a German zoologist wrote to say he'd been to New Guinea too and that the people living there had told him about a big animal with a long snout, which they referred to as a giant pig. It supposedly stood 6 feet tall at the shoulder, or 1.8 meters, and was very rare.

If you do a search for the devil-pig online, you'll see it called the gazeka in a lot of places. Let's discuss the word gazeka, because it doesn't have anything to do with New Guinea. In fact, it comes from an adaptation of a French musical called *The Little Michus*. I bet you didn't expect that. The musical is about two girls with the last name of Michu. One girl was given to the Michu family as a baby by her father, a general, who had to leave the country. The trouble is, the Michus had a baby daughter of the same age, and one day without thinking the father decided to give both babies a bath at the same time—and mixed them up. No one knew which girl was which, but they grew up as sisters who thought they were twins and were devoted to each other. The play takes place when they're both seventeen and the general suddenly shows up demanding his daughter back.

It's a funny musical and was popular in the original French in 1897, but in 1905 an English translation was performed in London and was a huge hit. It ran for 400 performances and became part of the pop culture of the day. So where does the gazeka come in?

George Graves was a famous English comic actor and he added an extra line or two to the play to get a laugh. He tells about a drunken explorer who thought he had seen a strange animal called the gazeka while under the influence of whiskey. The play was so popular, and the gazeka was considered so funny, that the idea just took off. The theater manager ran a competition for people to make drawings of the gazeka, and the winning drawing was made into a design that appeared on little charms, toys, and advertisements for sparkling water. The gazeka was even spun off into its own little song and dance in another play.

That was in 1905, remember. In spring of 1906 an explorer called Captain Charles A.W. Monckton led an expedition to Papua New Guinea, and on May 10 two members of the team were sent to investigate some tracks the expedition had found the previous day. The team members included an army private named Ogi and a village constable called Oina who acted as Ogi's guide. The two became separated at some point, and while he was looking for Oina, Ogi stumbled across two weird animals grazing in a grassy clearing. The devil-pigs!

The animals were only sort of pig-like. Later Ogi reported that they were dark in color with a patterned coat, cloven hooves, horse-like tail, and a long snout. They stood about 3.5 feet tall, or 106 centimeters, and were 5 feet long, or 1.5 meters. He shot at one but missed, probably because he was so scared, but he claimed later that his hands were shaking because he was cold.

The tracks the two men were investigating were of a large cloven-footed animal. Captain Monckton thought the tracks must be made by the devil-pigs.

The story hit the newspapers while the gazeka craze was still popular and people started calling the devil-pigs Monckton's Gazeka. Monckton didn't appreciate this, because he didn't like being compared to someone who saw imaginary animals while drunk.

So what could the devil-pig actually be?

The best guess is that it was an unknown species of tapir. We talked about the tapir in the water elephant chapter. The tapir looks kind of like a pig but with a short trunk called a proboscis that it mostly uses to gather plants. The proboscis is longer and much different than a pig's snout.

As far as we know, there have never been any tapirs in New Guinea. The only tapir that lives in Asia today is the Asian tapir, which is white with black forequarters and legs. It lives in lowland rainforests in Thailand, Sumatra, Myanmar, and a few other places, but not New Guinea. It's the largest species of tapir alive today, up to 3 feet 7 inches tall, or 110 centimeters.

In 1962 some stone carvings were discovered in Papua New Guinea. The carvings are a few thousand years old and depict a strange animal. It looks a little like an anteater sitting up on its bottom with its front paws on its round belly, although there's no tail. Its ears are small, its eyes are large, and it has a long nose with large nostrils at the end. It's usually said to depict the long-beaked echidna, a small spiny monotreme mammal that lives in New Guinea, although it doesn't look a lot like one.

In 1987 a mammologist named James Menzies looked at the carvings and made a suggestion. Instead of an echidna, he thought the carvings might depict a marsupial called a palorchestid diprotodont. Palorchestes is a genus of marsupials, with the largest species being about the size of a horse. It had large claws on the front feet and a long tongue like a giraffe's. Until recently, it was thought to have a short proboscis like a tapir, but a June 2020 study indicates it probably had prehensile lips instead. It used all these adaptations to strip leaves from branches.

Since Palorchestes probably didn't have a trunk after all, and since its fossil remains have only been found in Australia, and since it went extinct around 13,000 years ago, the carvings probably don't depict it. They probably also don't depict a tapir, although that's less certain. New Guinea is close to Australia and all of its native mammals are marsupials. The tapir is a placental mammal. That doesn't mean a species of tapir didn't once live on the island, but we have no fossil remains and the carvings don't resemble a tapir all that much.

One animal that definitely lives in New Guinea is the pig, which was introduced to the island thousands of years ago by humans. Wild boars might be responsible for the huge cloven hoof prints found by explorers in New Guinea.

That doesn't mean there isn't an unknown hoofed animal hiding on the island, though. New Guinea is still not very well explored by scientists or even locals, so there are certainly animals living there that are completely unknown to science. Maybe one is a giant tapir...or some other, more mysterious animal.

BEAST OF BUNGAY

On August 4, 1577, around mid-morning, a massive thunderstorm rolled through Suffolk, England. The Reverend Abraham Fleming later wrote an account of a bizarre event that happened during the storm.

It was a Sunday and church services were underway when the storm hit. During the lightning and thunder and torrential rain, Fleming wrote that a huge black dog entered St. Mary's Church in the small town of Bungay. It was clearly not an ordinary dog. Fleming wrote, in slightly edited modern English:

> This black dog, or the devil in such a likeness running all along down the body of the church with great swiftness, and incredible haste, among the people, in a visible form and shape, passed between two persons as they were kneeling in prayer and wrung the necks of them both at one instant clean backward, insomuch that even at a moment where they kneeled, they strangely died.

The dog also grabbed another man, resulting in the man appearing "drawn together and shrunk up, as it were a piece of leather scorched in a hot fire: or as the mouth of a purse or bag drawn together with a string." That man apparently recovered. The first two died.

But that's not all. Less than 10 miles away, or 16 kilometers, the storm advanced through the town of Blythburgh. In the Holy Trinity church the dog appeared again:

> The like thing entered, in the same shape and similitude, where placing himself on a main baulk or beam whereon sometime the rood did stand, suddenly he gave a swing down through the church, and there also, as before, slew two men and a lad, and burned the hand of another person that was among the rest of the company, of whom diverse were blasted. This mischief thus wrought, he flew with wonderful force to no little fear of the assembly, out of the church in a hideous and hellish likeness.

Fleming published his account in a pamphlet only a few weeks after the event took place, but he wasn't a witness. He also made some mistakes. He said that the two men who died after the dog wrung their necks backwards had been kneeling in prayer, but according to the parish register, both men who died had been in the belfry during the storm. Fleming also said that the dog left burnt claw marks on the door into St. Mary's church when it was actually the Holy Trinity church that was damaged. The church still has the same door and it's supposed to still show the claw marks. The marks don't look much like claw marks to me, but it's definitely possible that they were caused by lightning.

Fleming's account was probably heavily fictionalized to sell copies of his pamphlet, but that doesn't stop it from being a wonderfully creepy story based on an event that did actually happen. There really was a massive storm on that date that damaged both churches and killed several people, but other contemporary accounts of the storm don't mention a dog.

The rumor of a black dog in the storm might have started because there was an actual pet dog in the church or just outside that was frightened by the thunder and ran around in the church. Back then dogs were allowed in church but they sometimes barked or started fighting other dogs, at which point they had to be put outside. Many churches employed a man called a dog whipper to put dogs out, sometimes by using a big pair of metal tongs called dog tongs to grab a fighting dog and drag it outside. Like gigantic salad tongs, but for dogs.

Written accounts of ghostly black dogs go back over a thousand years in the British Isles and parts of Europe. The dogs are sometimes described as the size of a calf or even a pony, with glowing red eyes and shaggy fur. The very first black dog report anyone knows of is from France, recorded in the year 856. It occurred in a church too. A black dog with red eyes appeared in the church and ran around the altar several times before disappearing.

One well-known black dog is the Black Shuck of East Anglia, which is in eastern England and includes both Norfolk and Suffolk. The Black Shuck is a big black dog, sometimes described as having eyes as big as saucers, and in a few reports as having a single red eye in the middle of its face. The Beast of Bungay is actually considered to be part of the Black Shuck legend. Sightings of the Black Shuck still occur in Bungay, Blythburgh, and other parts of East Anglia.

People always like to know *why* something is happening, and there are lots of reasons given as to why a black dog appears. An account recorded in 1983 says that a girl was murdered on a road and after that a phantom hound started to be seen there, while other stories say that the dog is waiting for its master, a fisherman who was lost at sea. A popular variation of this legend says that a dog drowned along with its two masters and all were found washed up on shore. Since no one knew who the people were, they were buried in separate churchyards and the dog's spirit travels ceaselessly between the two graves. Another legend says that a dog guarding a house was killed by wolves and that its spirit continues to guard the area. Another says the ghostly black dog guards a treasure, usually gold. But some stories just say it's a demon or the spirit of a wicked person who died.

Here are a few accounts, all taken from a fantastic website called *Shuckland*.

This first story is from 1968 in the town of Barnby. "George Beamish... was walking home one night and coming up to the Water Bars when he noticed a dog alongside him... He did not pay any special regard to the animal, then turned to speak to it. He looked and he saw it was no ordinary dog. It was big and black, but it had no head. He put his hand down to [touch] the animal, but it went clean through the dog...there was nothing there. He got the wind up and ran home..."

Many stories are similar, since most black dog accounts take place on a road or path. For instance, this one:

In the early years of World War Two I was stationed on an airfield at Oulton in Norfolk. Sometime in the Winter of '41-42 I was walking along from Aylsham to Oulton Street. The night was very cold but clear. I had just passed Blickling Hall on my right when to my surprise I suddenly saw a large black dog standing in the middle of the road some few feet from me. As I called to the dog a most peculiar feeling came upon me. The nearest description I can give is that it was a 'nervous tingling.' I advanced towards the animal but as I went forward the animal retreated but without moving its feet, almost as though it was a cardboard 'cut-out' being pulled away from me with strings. The dog's mouth was open but it made no sound. ... I stopped and the dog also ceased its backward motion. After regarding me for maybe ten seconds the animal just completely disappeared. By 'disappeared' I mean that it did not run away but literally 'disappeared.' The night was very clear and I had a good view over the paddocks to my left and right. I could see no dog.

Sometimes a witness reports that the black dog disappears through some obstacle like a wall or a closed gate: for instance, this report from Earsham that probably occurred around 1920 to a Mrs. Wilson's father when he was young. It was mid-December near midnight, a clear moonlit night but with snow on the ground.

As he approached the last of the first row of cottages known as Temple Bar, he said he became aware of a horrible cold tingling sensation all over, and the feeling that his hair was standing 'on end.' At this point, he saw a large dog, probably black, come walking through the fence of the big private house known as 'The Elms' on his right, cross the road in front of him, a few feet away, and disappear through the WALL of the Rectory opposite...he found there was no sign whatsoever of any footprints, or other marks on the fresh snow. At this point he panicked and ran fast as he could to my Granny's house in the main street... At that time my father had no knowledge whatsoever of local ghosts...

A sighting of a black dog is usually taken as a bad omen, but sometimes a black dog seems to help people. In around 1842 in Catfield in Norfolk, "[s]everal women were out one night gathering rushes, trespassing on the marshes near Catfield Hall, when they heard the keeper coming. Suddenly a large black dog appeared...and started chasing back and forth among them, whimpering. Finally one of the [women] realized it wanted them to follow it, and it led them across the worst part of the marsh to a footpath, then on to a main road and home. When they looked around for the dog, it had disappeared."

In the early 20th century in Bawburgh, a young man whose name is only reported as Mr. E. Ramsey was riding his bicycle home late on a moonlit night.

As he got near his home village he saw, sitting by the signpost, 'the biggest hound' that he'd ever seen, with eyes that 'shone like coals of fire.' Although nervous he passed the dog, but it didn't move. Putting on speed he went on by, but half a mile further on heard him approaching from behind, 'his paws beating the grit road.' ...[T]he dog...went by him, 'so close [he] could smell [it].' When it was well in front the dog stopped suddenly beside a spinney, and stood in the middle of the road facing him, looking aggressive. Mr. Ramsey stopped and dismounted in fear, looking around for someone to help him, keeping the cycle between him and the hedge. But just at that moment an unlit vehicle roared out of the spinney, 'careering from side to side,' and seemed to crash straight into the dog. Mr. Ramsey fell into the hedge with the cycle on top [of] him, as the vehicle rushed by so close, and away up the lane out of sight. As the witness picked himself up, he was amazed to see the dog still standing there, as he was sure it had been struck. ...[T]o his surprise it just turned, and vanished into thin air. Mr. Ramsey...considered that it had saved his life on that night, since, if HE had been where the dog was, he would now be dead.

Black dogs have many names besides Black Shuck, most of which are local terms for the local black dog. These include Hairy Jack, Shag, Skriker, Padfoot, the Yeth or Yell Hound, the Barghest, the Churchyard Beast, and Hateful Thing. These are all names from various parts of the UK but black dogs are encountered in other places too, including parts of Europe, parts of the United States, especially in New England, and in parts of Mexico and South America. In many European mythologies, dogs symbolize death and the underworld, which may have influenced the black dog legends.

It's certain that at least some reports of ghostly black dogs were actually encounters with ordinary dogs that happened to be black. Many animals that are active at night exhibit eyeshine as light reflects off the tapetum lucidum. This helps the animal see better in the dark. The color of a dog's eyeshine depends on what color its eyes are but also depends on how much zinc or riboflavin is present in the pigments of its eyes, how old the dog is, and what breed it is. A dog's eyes can shine white, green, yellow, blue, purple, orange, or red. Some dogs even have different colored eyes, so that one eye shines yellow but the other shines green, or some other combination. A big dog with a black or dark brown coat, which would look black at night, which also has orange or red eyeshine, might be mistaken for the Black Shuck when encountered on a road at night by someone who's already familiar with the local legends.

That doesn't explain the ghostly dogs that vanish into thin air or walk through walls, though. Don't ask me to explain those. I love a good ghost story and I'm just going to appreciate how spooky those accounts are without worrying too much about what the black dog really is.

GHOST COW

There's a town in central New Jersey called Griggstown, and the Griggstown Cow was a legend told in the area. On foggy nights or rainy days, it was said, a solitary hunter or hiker might see a ghost cow in the mist near a canal outside of town. Occasionally someone would take a picture of the ghost cow, but the photos were all blurry and no tracks or manure were ever found.

The legend persisted for thirty years until November 23, 2002, when someone called the park office to report that the Griggstown Cow was stuck in a muddy ravine near the canal.

Sure enough, it was. It was a real live Holstein bull that had been living wild for decades after the area dairy farms closed, but he was old now and wasn't strong enough to get out of the ditch. Rescuers managed to hoist him out and he was left lying on the grass to recover. After two days he still hadn't managed to stand, so the park brought in a veterinarian to examine him. Unfortunately it turned out he was in such poor condition that the vet euthanized him so he wouldn't suffer. He was buried in the park.

It's a sad ending, but a thirty-year-old cow has lived a good long life. And if the Griggstown ghost cow can turn out to be a real animal, maybe other mystery animals are real too.

SELECTED TERMINOLOGY

Aestivation (also sometimes spelled estivation): A state of dormancy that takes place during the summer.

BCE and CE: "Before common era" and "common era." Replaces BC and AD to make the terms more inclusive.

Bovid: A member of the family Bovidae, which includes cattle, sheep, goats, and antelopes.

Brachiation: A method of moving through trees by swinging from arm to arm, as practiced by some monkeys and apes.

Cetacean: A whale, dolphin, porpoise, or other closely related aquatic mammal.

Chitin: The tough material that makes up an arthropod's exoskeleton, a cephalopod's beak, and similar body parts, mostly of invertebrates.

Crepuscular: An animal that is most active during dawn and dusk.

Cryptid: A term referring to a mystery animal, especially one with supernatural attributes. This is not a term used by scientists.

Cryptozoology: The study of mystery animals. This is not a scientific field, just a way to describe a hobby. Zoologists, biologists, and other scientists seek out and study mystery animals and animal mysteries as a normal part of their work.

Diurnal: An animal that is most active during the day.

Hibernation: A state of dormancy that takes places during the winter.

Keratin: The tough material that makes up hair, feathers, claws, hooves, fingernails, horns, scales, and similar body parts, usually of vertebrates.

Nocturnal: An animal that is most active at night.

Osteoderm: A bony structure in the skin that acts as armor. Examples are the armadillo's shell and the hard plates beneath a crocodilian's skin.

Rostrum: A beaklike snout. The hard snout of a dolphin is a rostrum, for instance. So is the saw of a sawfish.

Ruminant: A hoofed animal that digests plant material using foregut fermentation. Ruminants chew their cud as part of the digestion process. Bovids are ruminants, as are deer and giraffes.

Scute: A bony scale on an animal's skin that acts as armor. Examples are a turtle's shell and the hard scales on top of a crocodilian's skin. Unlike actual scales, scutes do not overlap. Confusingly, the armadillo's shell can also be described as made up of scutes.

Subfossil: Remains of an animal that are too recent to be fully fossilized and are instead partially fossilized or unfossilized. They may contain organic matter that can be genetically tested. Most subfossil remains are less than half a million years old.

Ungulate: Odd-toed mammals, including horses and their relations, rhinoceroses, and tapirs. Ungulates digest plant material using hindgut fermentation and do not chew cud.

SELECTED SOURCES

Arment, Chad, editor. (2006). *Cryptozoology and the investigation of lesser-known mystery animals.* Coachwhip.

Bales, Lyn B. (2010). *Ghost birds: Jim Tanner and the quest for the ivory-billed woodpecker, 1935-1941.* Univ. of Tennessee.

Bartram, W. (2001). *Travels through North and South Carolina, Georgia, east & west Florida, the Cherokee country*, etc. Electronic ed., 151-152. Univ. of North Carolina at Chapel Hill. tinyurl.com/kxe97n98

Beebe, William. (1934). *Half mile down.* Harcourt, Brace & Co.

Bressan, David. (2012, October 7). "De Loys' ape." *Scientific American.* https://tinyurl.com/55a8xmwa

Cuninghame, R.J. (1912). "The water-elephant." *Journal of the East Africa and Uganda Natural History Society, 2*(4), 97-8.

Dash, Mike. (1994). "The devil's hoofmarks: Source material on the great Devon mystery of 1855." *Fortean Studies.* https://tinyurl.com/2ry399v2

Drinnon, Dale A. (2011). "Three kinds of unicorns." https://tinyurl.com/28xx6bfm

Fama, Elizabeth. (2012, Aug 16). "Debunking a great New England sea serpent." https://tinyurl.com/3fzmfr26

Forth, Gregory. (2014). "Gugu." *Anthropos* 109, 149-160. https://tinyurl.com/xscnff52

Forth, Gregory. (2012). *Images of the wildman in Southeast Asia: An anthropological perspective.* Routledge.

France, R.L. (2017). "Imaginary sea monsters and real environmental threats: reconsidering the famous *Osborne*, 'moha-moha', *Valhalla*, and 'Soay beast' sightings of unidentified marine objects." Int'l Review of Environmental History 3(1), 63-100.

Greenland, Felicity and Hayward, Philip. (2020, Sept. 23). "NINGEN: The generation of media-lore concerning a giant, sub-Antarctic, aquatic humanoid and its relation to Japanese whaling activity." *Shima: The International Journal of Research into Island Cultures.* 14(1), 133–151. https://tinyurl.com/4rjzhbtj

Haslam, Garth. (n.d.) "1954, August 11: The Canvey Island monsters." *Anomalies: The Strange and Unexplained.* https://tinyurl.com/dmah9r7s [retrieved 8/6/2021]

Heatherson, Liam. (2012, Nov. 11). "The Canvey Island monster: My findings." *CanveyIsland.org.* https://tinyurl.com/2p8afeyw [retrieved 12/29/2021]

Hidden East Anglia. (n.d.) *Shuckland.* https://tinyurl.com/u9u7ky7d

Izzard, Ralph. (1951). *The Hunt for the Buru.* Linden Publishing.

Lammertink, Martjan, et al. (2011, Oct 1). "Film documentation of the probably extinct imperial woodpecker (*Campephilus imperlalis*)." *Auk*, 128(4), 671-677. https://tinyurl.com/ycxm2r8v

Ley, Willy. (1968). *Dawn of zoology.* Prentice-Hall.

Ley, Willy. (1987). *Exotic zoology.* Bonanza. (Original work published 1959)

Ley, Willy. (1948). *The lungfish, the dodo, & the unicorn.* Viking.

Loxton, Daniel and Prothero, Donald R. (2013). *Abominable science! Origins of the yeti, Nessie, and other famous cryptids.* Columbia Univ.

Maruna, S. (2006). "Substantiating Audubon's Washington eagle." *Ohio Cardinal*, 29(3), 140-150. https://tinyurl.com/fnz67pb8

Meier, Allison C. (2018, Dec 4). "The strange nature of the first printed illustration of a sloth." *Smithsonian Magazine.* https://tinyurl.com/4z93th6r

Naish, Darren. (2017, Feb 3). "Animal species named from photos." *Tetrapod Zoology.* https://tinyurl.com/dsrf4azu [retrieved 8/22/2021]

Naish, Darren. (2017). *Hunting monsters: Cryptozoology and the reality behind the myths.* Arcturus.

Naish, Darren. (2013, Nov. 23). "The amazing Hook Island sea monster photos, revisited." *Tetrapod Zoology*. https://tinyurl.com/2p88pn4h [retrieved 12/29/2021]

Sharma, R., Goossens, B., Heller, R. *et al.* (2018). "Genetic analyses favour an ancient and natural origin of elephants on Borneo." *Sci Rep* 8, 880. https://tinyurl.com/473kd6ww

Shuker, Karl P.N. (2007). *Extraordinary animals revisited*. CFZ Press.

Shuker, Karl P.N. (2021, April 24). "How the Nandi bear was conclusively identified and contentiously lost." *ShukerNature*. https://tinyurl.com/4a9xeeav

Shuker, Karl P.N. (2003). *The search for the last undiscovered animals*. Fall River.

Simons, Elwyn L. (2010). "Inferences about the distant past in Madagascar." *Lemur News* 15, 25-27. https://tinyurl.com/f92fv2de

Snyder, N., & Fry, J. (2013). "Validity of Bartram's painted vulture (Aves: Cathartidae)." Zootaxa, 3613(1), 61–82. https://tinyurl.com/y822pbu2

Soini, Wayne. (2010). *Gloucester's Sea Serpent*. History Press.

Steller, Georg W. (1988). *Journal of a voyage with Bering, 1741-1742*. (M.A. Engel and O.W. Frost, translators). Stanford Univ. Press.

Strauss, Bob. (2020, Aug 28). The mystery of North America's black wolves. https://tinyurl.com/4h3akhm5

Turnbo, S.C. (1844-1925). *The Turnbo Manuscripts*. https://tinyurl.com/423bwzaf

White, T.H., translator. (1984). *The book of beasts: Being a translation from a Latin bestiary of the twelfth century*. Dover. (Original work published 1954)

Whittaker, Tony. (1990). "Kawekaweau – myth or monster?" *New Zealand Geographic*. https://tinyurl.com/5ka34964

Whittall, Austin. (2012). *Monsters of Patagonia*. Zagier & Urruty.

ALPHABETICAL INDEX

SPECIAL THANKS

Thanks so much to everyone who contributed to the Kickstarter! I couldn't have done it without all of you.

Netherworld Post Office and Lee Raye and Matt Krish and Zoe & Dillon Husock and Emily Beaman and Robert "Midwestern Tanuki" Sroka and Sergey Kochergan and Steven Gernandt and Mel McMeans and Jesse Grier and Adam Eaton and Karen Koy and Dryden and Derek Shafer and Megan and Paul Csomo and Isaac Sills and Tara Moore and chris lentil and Declan Morrison, Finn Morrison, Emry Morrison and Bradley Johnson and Dan Kuhlman and Simon Woolley and Jake & Lilly Burns and Ann Middleton and Carolin Rother and Blake Smith and Michael Miele and Grace and Seamus Hudson and Ava Jarvis and Miranda Forner and Megan Engelhardt and Giovanna B. Derhofer and Connor Roessler and Cristin Chall and Catie L. and Sarah Liberman and Aaron Dunn and Caitie Sith and David Eccher and Stephany Spencer and Joyce Sully and H Lynnea Johnson and Nicholas and Darwin the beagle and Nicola and Pip and Clay Henry Corwin and Jason & Kristie Follett and Joanne B Burrows and Alina Osterlund and Kevin Eldridge and Lord Bob and Jasper, Ed and Brooke Love and Robert Plant and Heather & Casey Reuck and Emery and Leete Morton and Derek Wheaton and Autumn Dominguez and Arthina Selchow and Max & Ellie Brandon and Zachary Thomas Tyler and Yori and Chloe and Gadi and Mary and Jillian and Kristjan Wager and Jennifer and John and Keith and Heather and Hari and Juliano and Sara and Hailie and my lovely and charming cousin Molly and my devoted aunt, the beautiful brainiac Janice McClelland!

Thanks also to the Strange Animals Podcast Patreon subscribers, past and present (and future).

Thanks to everyone who's left a review or shared the podcast or Kickstarter information on social media or with friends.

Thanks to everyone who's supported me with kind words, corrections, topic suggestions, and links to cool animal videos.

Thanks to Blake Smith of *MonsterTalk* (and other awesome projects), who was enthusiastic about the book before it was even completely finished and wrote the foreword!

And extra special thanks to my other animal podcasters! I'd give you all shout-outs but I'd hate to leave anyone out by accident.

Podcasting can be a solitary hobby but you've all made me feel part of a weird but awesome family.

NOTES

ELENGASSEN

1. Whittall, 2012
2. Whittall, 2012

PAINTED VULTURE

1. Bartram, 2001
2. Snyder, 2013

BURU

1. Izzard, 1951

BEEBE'S DEEP-SEA MYSTERY FISH

1. Beebe, 1934

STELLER'S SEA-APE

1. Steller, 1988
2. Steller, 1988

GIANT CENTIPEDES

1. Turnbo

DE LOYS' APE

1. Tejera, quoted in Bressan, 2012

NOTES

DEVIL'S FOOTPRINTS

1. Leutscher, quoted in Dash, 1994

www.ingramcontent.com/pod-product-compliance
Lightning Source LLC
Chambersburg PA
CBHW062113020426
42335CB00013B/947